The Secret of the Sea

THE SECRET OF THE SEA

An Anthology of Underwater Exploration and Adventure

COMPILED BY

Dorothy Clewes

ILLUSTRATED BY

Jeroo Roy

HEINEMANN : LONDON

William Heinemann Ltd
15 Queen Street, Mayfair, London W1X 8BE

LONDON MELBOURNE TORONTO
JOHANNESBURG AUCKLAND

First published 1973

434 93286 8

Filmset and printed in Great Britain by
BAS Printers Limited, Wallop, Hampshire

Contents

'Wouldst thou' – so the helmsman answered, –
'Learn the secrets of the sea?
Only those who brave its dangers
Comprehend its mystery!'

Longfellow

Preface

The ocean covers three-fourths of the earth's surface. Men have dived and swum and fished in it from earliest times. Poets have woven epic stories from the adventures and tragedies that have taken place on or beneath the dark impenetrable waters, and novelists have let their imagination run riot – some of them prophetically – on the mysteries it might hold. Off the pearl coasts of Ceylon, the Philippine Islands, New Guinea and Borneo, and some coasts of Australia, men and women have shortened their lives diving for the elusive pearl. Using only control of their breath to sustain them for their brief stay underwater, they risk the crippling pressure of the water for the rewards that the finding of pearls will bring them. Without apparatus the time that can be spent beneath its surface is minimal, and so men worked to find some way of overcoming this hindrance. The invention of the air pump was a great step forward and was followed by that of the diving-bell – a bell-shaped tank which was lowered into the sea, in which sufficient air pressure was maintained to repel the water, thus enabling men to work on the sea-bed for long periods. This was followed by the pressurized rubber diving suit, developed from that invented by Auguste Siebe in the early nineteenth century. It consisted of a one-piece suit, weighted with lead, with a metal breast-plate and a helmet to which an airline was attached. The air pressure within the flexible suit had to equal the external water pressure, but there are limits to the pressure the human body can withstand, so that such apparatus was serviceable only in comparatively shallow water.

The ocean is by far the largest part of our environment and yet we are only of recent years beginning to study it in all its aspects. We fish in it, we dive in it, we swim in it – but only in this century have we started really to explore it. We have many men to thank for this new awakening, trail-blazers all of them, and in the forefront was William Beebe, who, working with Mr Otis Barton, came up with the first really successful submarine capsule. There had been many earlier unsuccessful attempts – even in the centuries before Christ – as man's insatiable curiosity drove him to probe this most fascinating of all elements, but William Beebe's bathysphere really worked. In its small, confined space he was able to descend half a mile into a veritable wonderland. Professor Piccard followed on with a more elaborate device which he called a bathyscaph. This, in shape, was not unlike the balloon his father, Auguste Piccard, had invented in order to study cosmic rays in the stratosphere, only Jacques Piccard's 'balloon' was to operate in another element: the sea. It was a submarine for taking men to the unexplored depths where no man had ever been before. It had its own buoyancy and sinking devices, and with its own supply of air, was free of any attendant pipeline to the surface.

Then came a breakthrough. In 1943, Captain Jacques-Yves Cousteau invented and perfected the first 'aqualung'. This consisted of three moderate-sized cylinders of compressed air, linked to an air regulator the size of an alarm-clock. The bottles contained air condensed to one hundred and fifty times atmospheric pressure. From the regulator there extended two tubes, joining on to a mouthpiece. With the equipment harnessed to the back, a watertight glass mask over eyes and nose, and rubber foot fins, unencumbered flights could be made in the depths of the sea.

It was as if the key had been found to a long-closed door. Biologists were now able to study marine life, archaeologists to work on underwater wrecks and submerged sites, and ordinary men and women could go underwater hunting, or just dive for the sheer pleasure of exploring this new and exciting element. If one wanted to cruise only just beneath the surface a snorkel, mask, and fins was all that was needed: rocks and submerged coral reefs teemed with life and interest; fish, some curious, all of them seemingly unafraid, appeared to accept this new fellow creature in their midst with little surprise. For deeper diving, the addition of the air containers made it possible to stay down for as long as the air lasted. Diving clubs began to spring up, all with differing aims in mind. Cameras encased in watertight boxes could be taken down: it was found that it was no more difficult to take pictures underwater than anywhere else. Jacques Cousteau and Hans Hass – to name only two professional divers whose names are now household words – began to put their knowledge and expertise to good use, making wonderful films of this underwater world, whetting the appetites of thousands of men and women to emulate them, to experience for themselves the thrill and excitement of so new and exotic a pastime.

One of the more obvious and romantic aspects of diving is the search for and the discovery of foundered ships. New electronic equipment has suddenly made this much easier. Echo sounders can be trawled over a suspected site and the image of the object is automatically drawn for you. Now piracy on the high seas is matched by piracy under them. A wreck having been legitimately acquired can be plundered by rival groups without there being any legal redress. This is a situation which must – and will be – put to rights. Mr Sydney Wignall, who was given an award for his work on Spanish Armada vessels, has gone a long way to help bring this about after the pirating experiences he had off the West Coast of Ireland while excavating the wreck of the *Santa Maria de la*

Rosa which went down off the Blaskets. Ships of the Spanish Armada are scattered on the sea floor from Spain to the English Channel and all around our coast. Some are in known, marked positions, others went down unobserved in the terrible storms they encountered, or foundered on the cruel and hidden rocks that abound among these islands. They lie there on their burial grounds waiting for the unsuspecting and intrepid diver who may one day come across the unnatural form on the seabed and recognize it for what it really is. Not an easy achievement when centuries of submersion have put them at the mercy of changing tides and marine organisms. A wealth of history is there for the discovering, a glimpse of a bygone civilization in miniature.

Helmet divers have always played their unsung part in underwater salvage, in peacetime and in war, removing mines laid by an enemy, or removing mines laid by ourselves in an endeavour to defend ourselves against an enemy; in underwater repair work; in rescue diving, the laying of telephonic communications; the throwing of bridges across rivers; and more recently, the building of oil rigs in mid-ocean. Presumably the helmet will always have its part to play – but the freedom that the aqualung has given has revolutionized the scope for previously unthought of activities.

In this anthology I have tried to follow, in fact, fiction and verse, the history of underwater activity, and the fascination it has for all of us, whether we just want to read about it, or take an active part in it. More than that, I hope I have made you want to read in their entirety the books out of which I have taken such small excerpts, the writers of which have been, and still are, many of them, trailblazers themselves into a world that is as yet unspoiled and beautiful.

The Forsaken Merman

Mathew Arnold

Come, dear children, let us away;
Down and away below.
Now, my brothers, call from the bay;
Now the great winds shoreward blow;
Now the salt tides seaward flow;
Now the wild white horses play,
Champ and chafe, and toss in the spray.
Children, dear, let us away.
This way, this way!

Call her once before you go.
Call her once yet.
In a voice that she will know:
'Margaret! Margaret!'
Children's voices should be dear
(Call once more) to a mother's ear;
Children's voices wild with pain.
Surely she will come again.
Call her once and come away,
This way, this way!
'Mother, dear, we cannot stay.'
The wild white horses foam and fret.
Margaret! Margaret!

Come, dear children, come away down.
Call no more.
One last look at the white-walled town,
And the little grey church on the windy shore
Then come down.
She will not come though you call all day.
Come away, come away.

Children, dear, was it yesterday
We heard the sweet bells over the bay?
In the caverns where we lay,
Through the surf and through the swell,
The far-off sound of a silver bell?
Sand-strewn caverns, cool and deep,
Where the winds are all asleep;
Where the spent lights quiver and gleam;
Where the salt weed sways in the stream;
Where the sea-beasts, ranged all round,
Feed in the ooze of their pasture-ground;
Where the sea-snakes coil and twine,
Dry their mail, and bask in the brine;
Where great whales come sailing by,
Sail and sail, with unshut eye,
Round the world for ever and aye?
When did music come this way?
Children, dear, was it yesterday?

Children, dear, was it yesterday
(Call yet once) that she went away?
Once she sat with you and me,
On a red gold throne in the heart of the sea,
And the youngest sat on her knee.
She combed its bright hair, and she tended it well,
When down swung the sound of a far-off bell.
She sighed, she looked up through the clear green sea.

She said: 'I must go for my kinsfolk pray
In the little grey church on the shore today.
'Twill be Easter-time in the world – ah, me!
And I lose my poor soul, Merman, here with thee.
I said: Go up, dear heart, through the waves.
Say thy prayer, and come back to the kind sea-caves.
She smiled, she went up through the surf in the bay.
Children, dear, was it yesterday?

Children, dear, were we long alone?
'The sea grows stormy, the little ones moan.
Long prayers,' I said, 'in the world they say.
Come,' I said, and we rose through the surf in the bay.
We went up the beach, by the sandy down
Where the sea-stocks bloom, to the white-walled town.
Through the narrow-paved streets, where all was still,
To the little grey church on the windy hill.
From the church came a murmur of folk at their prayers,
But we stood without in the cold-blowing airs.
We climbed on the graves, on the stones worn with rains,
And we gazed up the aisle through the small leaded panes.
She sate by the pillar; we saw her clear;
'Margaret, hist! Come quick, we are here.
Dear heart,' I said, 'we are long alone.
The sea grows stormy, the little ones moan.'
But, ah! she gave me never a look,
For her eyes were sealed to the holy book.
Loud prays the priest; shut stands the door.
Come away, children, call no more.
Come away, come down, call no more.

Down, down, down;
Down to the depths of the sea.
She sits at her wheel in the humming town,
Singing most joyfully.

Hark what she sings: 'O joy, O joy,
For the humming street and the child with its toy,
For the priest and the bell, and the holy well,
For the wheel where I spin,
And the blessed light of the sun.'
And so she sings her fill,
Singing most joyfully,
Till the shuttle falls from her hand,
And the whizzing wheel stands still.

She steals to the window, and looks at the sand;
And over the sand at the sea;
And her eyes are set in a stare,
And anon there breaks a sigh,
And anon there drops a tear,
From a sorrow-clouded eye,
And a heart sorrow-laden,
A long, long sigh,
For the cold, strange eyes of a little Mermaiden,
And the gleam of her golden hair.

Come away, away children,
Come children, come down;
The hoarse wind blows colder;
Lights shine in the town.
She will start from her slumber
When gusts shake the door;
She will hear the winds howling,
Will hear the waves roar.
We shall see while above us
The waves roar and whirl,
A ceiling of amber,
A pavement of pearl.
Singing: 'Here came a mortal,
But faithless was she;

And alone dwell for ever
The kings of the sea.'

But, children, at midnight,
When soft the winds blow;
When clear falls the moonlight,
When spring-tides are low;
When sweet airs come seaward
From heaths starred with broom,
And high rocks throw mildly
On the blanched sands a gloom;
Up the still, glistening beaches,
Up the creeks we will hie;
Over banks of bright seaweed
The ebb-tide leaves dry.
We will gaze from the sandhills
At the white, sleeping town,
At the church on the hillside –
And then come back down.
Singing: 'There dwells a loved one,
But cruel is she;
She left lonely for ever
The kings of the sea.'

Jonah and the Whale
from
The Bible Designed to be Read as Literature

Is this story a myth or is it true? One has to be very careful in these scientific days of coming to a snap decision on almost anything. Once, not so very long ago, we would have answered without hesitation that it was a story – or maybe a parable, as so many of the Bible storys are, intent as they are on pointing out to us some wise philosophy. But now we have more knowledge of the sea and of the strange and wonderful creatures who inhabit this world. Brave and curious men have gone down into its depths to see for themselves what this underwater world is like and to get to know something of the life that abounds there. Hans Hass, who is one of the great names in the early history of underwater diving, was at one time a visitor to a whaling station at Sao Vincente in the Azores where he and his friends watched a whale carcass being cut up. He said: 'We had a great surprise when the belly of the whale – some fifty feet long – was opened. In it were two half-digested sharks. One was two and a half metres long and the other was three metres and ten centimetres. The legend of Jonah assumed a new significance for us. A big sperm whale would be quite capable of swallowing a man whole.'

So, perhaps we should read the story with greater respect and not be so sure that we know it all, because knowledge is never static, it changes with our growing awareness and experience. Things we do not believe today we may be forced to admit to tomorrow.

Now the word of the Lord came unto Jonah the son of Amittai, saying: 'Arise, go to Nineveh, that great city, and cry against it; for their wickedness is come up before me'.

But Jonah rose up to flee into Tarshish from the presence of the Lord, and went down to Joppa; and he found a ship going to Tarshish: so he paid the fare thereof, and went down into it, to go with them unto Tarshish from the presence of the Lord.

But the Lord sent out a great wind into the sea, and there was a mighty tempest in the sea, so that the ship was like to be broken. Then the mariners were afraid, and cried every man until his god, and cast forth the wares that were in the ship into the sea, to lighten it of them. But Jonah was gone down into the sides of the ship; and he lay, and was fast asleep.

So the shipmaster came to him, and said unto him: 'What meanest thou, O sleeper? Arise, call upon thy God, if so be that God will think upon us, that we perish not.'

And they said every one to his fellow: 'Come and let us cast lots, that we know for whose cause this evil is upon us.' So they cast lots, and the lot fell upon Jonah.

Then said they unto him: 'Tell us, we pray thee, for whose cause this evil is upon us; what is thine occupation? and whence comest thou? what is thy country? and of what people art thou?'

And he said unto them: 'I am a Hebrew; and I fear the Lord, the God of heaven, which hath made the sea and the dry land.'

Then were the men exceedingly afraid, and said unto him: 'Why hast thou done this?' For the men knew that he fled from the presence of the Lord, because he had told them. Then said they unto him: 'What shall we do unto thee, that the sea may be calm unto us?' for the sea wrought, and was tempestuous.

And he said unto them: 'Take me up, and cast me forth into the sea; so shall the sea be calm unto you; for I know that for my sake this great tempest is upon you.'

Nevertheless the men rowed hard to bring it to the land; but they could not: for the sea wrought, and was tempestuous against them. Wherefore they cried unto the Lord, and said: 'We beseech thee, O Lord, we beseech thee, let us not perish for this man's life, and lay not upon us innocent blood: for thou, O Lord, hast done as it pleased thee.'

So they took up Jonah, and cast him forth into the sea: and the sea ceased from her raging. Then the men feared the Lord exceedingly, and offered a sacrifice unto the Lord, and made vows.

Now the Lord had prepared a great fish to swallow up Jonah. And Jonah was in the belly of the fish three days and three nights. Then Jonah prayed unto the Lord his God out of the fish's belly and said:

'I cried by reason of mine affliction unto the Lord,
And he heard me;
Out of the belly of hell cried I,
And thou heardest my voice.
For thou had cast me into the deep,
In the midst of the seas;
And the floods compassed me about:
All thy billows and thy waves passed over me.
Then I said: 'I am cast out of thy sight;
Yet I will look again toward thy holy temple.'
The waters compassed me about, even to the soul:
The depth closed me round about,
The weeds were wrapped about my head.
I went down to the bottoms of the mountains;
The earth with her bars was about me forever.
Yet hast thou brought up my life for corruption, O Lord my God.
When my soul fainted within me I remembered the Lord:
And my prayer came in unto thee,
Into thine holy temple.
They that observe lying vanities forsake their own mercy.
But I will sacrifice unto thee with the voice of thanksgiving:
I will pay that that I have vowed.
Salvation is of the Lord.'

And the Lord spake unto the fish, and it vomited out Jonah upon the dry land.

Twenty Thousand Leagues Under the Sea

Jules Verne

Jules Verne was a French novelist who lived between the years 1828–1905 and in that time achieved great and enduring popularity with his books, which were a combination of adventure and popular science. In 1873 he wrote this story about M. Pierre Aronnax and his faithful friend and servant, Conseil, who, together with the famous Canadian harpooner, Ned Land, set out to solve the mystery of – and destroy if possible – the sea monster that has attacked and caused the deaths of so many innocent travellers on the oceans of the world. They are washed overboard from the frigate in which they are stalking the monster, and snatched from certain death by the very beast they hoped to overthrow. But the beast turns out to be the fantastic submarine, the *Nautilus*, whose commander, Captain Nemo, after the loss at the hands of an enemy of all he had ever loved has denounced life on this earth and vowed terrible vengeance on his oppressors. Jules Verne's great imaginative vision gives us a startling and prophetic look into the future. This is a very condensed version taken from *The Children's Encyclopaedia*, and perforce much has been left out. For the full, spine-chilling effect the book must be read in its entirety.

In the year 1866 the whole seafaring world of Europe and America was greatly disturbed by an ocean mystery which baffled the wits of scientists and sailors alike. Several vessels, in widely different regions of the seas, had met a long and rapidly moving object, much larger than a whale, and capable of almost incredible speed. It had also been seen at night, and was then phosphorescent, moving under the water in a glow of light.

There was no doubt whatever as to the reality of this unknown terror of the deep, for several vessels had been struck by it, and particularly the Cunard steamer *Scotia*, homeward bound for Liverpool. It had pierced a large triangular hole through the steel plates of the *Scotia's* hull and would certainly have sunk the vessel had it not been divided into seven watertight compartments, any one of which would stand injury without danger to the vessel. It was three hundred miles off Cape Clear that the *Scotia* encountered this mysterious monster. Arriving after some days' delay at Liverpool, the vessel was put into dock, when the result of the blow from the unknown was thoroughly investigated. So many vessels having recently been lost from unknown causes, the narrow escape of the *Scotia* directed fresh attention to this ocean mystery, and both in Europe and America there was a strong public agitation for an expedition to be sent out, prepared to do battle with, and if possible to destroy, this narwhal of monstrous growth, as many scientists believed it to be.

Now I, Pierre Arronax, assistant professor in the Paris Museum of Natural History, was at this time in America, where I had been engaged on a scientific expedition into the disagreeable region of Nebraska. I had arrived at New York in the company of my faithful attendant, Conseil, and was devoting my attention to classifying the numerous specimens I had gathered for the Paris Museum. As I had already some reputation in the scientific world from my book on *The Mysteries of the Great Submarine Grounds*, a number of people did me the honour of consulting me concerning the one subject then exercising the minds of all interested in ocean travel.

An expedition was also being fitted out by the United States government, the fastest frigate of the navy, the *Abraham Lincoln*, under command of Captain Farragut, being in active preparation, with the object of hunting out this wandering monster which had last been seen three weeks before by a San Francisco steamer in the North Pacific Ocean. I was invited to join this expedition as a representative of France, and immediately decided to do so. The faithful Conseil said he would go with me wherever I went, and thus it came about that my sturdy Flemish companion, who had accompanied me on scientific expeditions for ten years, was with me again on the eventful cruise which had begun when we sailed from Brooklyn for the Pacific and the unknown.

The crew of the frigate and the various scientists on board were all eagerness to meet the great cetacean, or sea-unicorn. My own opinion was that it would be found to be a narwhal of monstrous growth, for these creatures are armed with a kind of ivory sword, or tusk, as hard as steel, and sometimes almost seven feet long, by fifteen inches diameter at the base. Supposing one to exist ten times as large as any that had ever been captured, with its tusk proportionately powerful, it was conceivable that such a gigantic creature, moving at a great rate, could do all the damage that had been reported.

There was among our crew one Ned Land, a gigantic Canadian of forty who was considered to be the prince of harpooners. Many a whale had received its death-blow from him, and he was eager to flesh his harpoon in this redoubtable cetacean which had terrified the marine world.

Week after week passed without any sign that our quest would be successful. Indeed, after nearly four months had gone, and we had explored the whole of the Japanese and Chinese coasts, the Captain reached the point of deciding to return, when one night the voice of Ned Land was heard calling:

'Look out there! The thing we are looking for on our weather-beam!'

At this cry the entire crew rushed towards the harpooner – captain, officers, masters, sailors, and cabin-boys; even the engineers left their engines, and the stokers their furnaces. The frigate was now moving only by her own momentum, for the engines had been stopped.

My heart beat violently. I was sure the harpooner's eyes had not deceived him. Soon we could all see, about two cables' length away, a strange and luminous object, lying some fathoms below the surface, just as described in many of the reports. One of the officers suggested that it was merely an enormous mass of phosphorous particles, but I replied with conviction that the light was electric. And even as I spoke the strange thing began to move towards us!

The captain immediately reversed engines and put on full speed, but the luminous monster gained on us and played around the frigate with frightful rapidity. Its light would go out suddenly and reappear again on the other side of the vessel. It was clearly too great a risk to attack the thing in the dark, and by midnight it disappeared, dying out like a huge glow-worm. It appeared again about five miles to the windward, at two in the morning, coming up to the surface as if to breathe, and it seemed as if the air rushed into its huge lungs like steam in the vast cylinders of a 2,000 horse-power engine.

'Hm,' said I. 'A whale with the strength of a cavalry regiment would be a pretty whale!'

Everything was in readiness to attack with the coming of dawn, and Ned Land was calmly sharpening his great harpoon, but by six in the morning the thing had again disappeared, and a thick sea-fog made it impossible to observe its further movements. At eight o'clock, however, the mist had begun to clear, and then, as suddenly as on the night before, Ned Land's voice was heard calling: 'The thing on the port-quarter!'

There it was, surely enough, a mile and a half away, now a large black body showing above the waves, and leaving a track of dazzling white as its great tail beat the water into foam.

Moving rapidly it approached within twenty feet of the

frigate. Ned stood ready at the bow to hurl his harpoon and the monster was now shining again with that strange light which dazzled our eyes. All at once he threw the harpoon. It struck on a hard body. Instantly the light went out and two enormous water-spouts fell on our deck. A frightful shock followed, and the next moment I found myself struggling in the sea. Though a good swimmer I kept afloat with some difficulty, and great was my joy when I heard the voice of the faithful Conseil, who had jumped in after me. Much stronger than myself, he helped me to remove some of my clothing and thus we kept afloat until I fainted.

When I regained consciousness, I found myself on the top of what seemed to be a floating island, and there was Ned Land as well as Conseil. We were on the back of the mysterious monster, and it was made of metal! Presently it began to move, and we were afraid it might go below the surface.

Indeed, it seemed to be on the point of submerging, when Land hammered loudly on the metal plates, and in a moment an opening was made and the three of us were drawn inside by eight masked men. A door banged on us, and for half an hour we lay in utter darkness. Then a brilliant electric light flooded the cabin, a room of about twenty feet by ten, and two men entered. One was tall, pale and dark-eyed, but magnificently proportioned.

Though we spoke to them in French, German, English, and Latin, they did not seem to understand, while their own speech was unintelligible to us. But they gave us clothes and food. After eating the food, which was strange but delicious, we all lay down and slept the sleep of sheer exhaustion.

Next day the tall man, whom I afterwards came to know as Captain Nemo, master of this marvellous submarine boat, came to me and speaking in French, said:

'I have been considering your case, and did not choose to speak until I had weighed it well. You have pursued me to destroy me. I have done with society for reasons of my own. I have decided. I give you choice of life or death. If you grant me a passive obedience, and submit to my consigning you to your cabin for some hours or days, as occasion calls, you are safe. You, Monsieur Arronax, have least cause to complain, for you have written on the life of the sea – I have your book in my library here – and you will benefit most when I show you its marvels. I love it. It does not belong to despots.'

Clearly we could do nothing but submit, and afterwards Captain Nemo showed me his wondrous craft.

It was indeed a thing of marvels; for, besides the dining-room, it contained a large library of ten thousand volumes, a drawing-room measuring thirty feet by eighteen, and fifteen high. The walls of this apartment were adorned with masterpieces of the great painters, and beautiful marbles and bronzes. A large piano-organ stood in one corner, and there were glass cases containing the rarest marine curiosities which a naturalist could hope to see. A collection of enormous pearls in a cabinet must have been worth millions, and Captain Nemo told me he had rifled every sea to find them.

The room assigned to me was fitted up with every luxury, yet the captain's own apartment was as simply furnished as a monastic cell, but in it were contained all the ingenious instruments that controlled the movements of the *Nautilus*, as his submarine was named. The electricity was manufactured by a process of extracting chloride of sodium from the sea-water, but the fresh air necessary for the life of the crew could only be obtained by rising to the surface. The engine-room was sixty feet long, and in it was the machinery for producing electricity as well as that for applying the power to the propeller.

The *Nautilus*, Captain Nemo told me, was capable of a speed of fifty miles an hour, and could be made to sink or rise with precision by flooding or emptying a reservoir. In a box, raised somewhat above the hull and fitted with glass ten inches thick, the steersman had his place, and a powerful electric reflector behind him illumined the sea for half a mile in front.

The submarine also carried a small torpedo-like boat, fitted in a groove along the top, so that it could be entered from the *Nautilus* by opening a panel, and, after that was closed, the boat could be detached from the submarine and would then bob upwards to the surface like a cork. The importance of this and its bearing on my story will appear in due time.

It was on a desert island that Captain Nemo had carried out the building of the *Nautilus*, and from many different places he had secured the various parts of the hull and machinery, in order to maintain secrecy.

Deeply interested as I was in every detail of this extraordinary vessel, and excited beyond measure at the wonders which awaited me in exploring the world beneath the waves, I had still the feeling of a prisoner who dared scarcely hope that liberty might some day be obtained. But when the metal plates which covered the windows of the saloon were rolled back as we sailed under the water, and on each hand I could see a thronging army of many-coloured aquatic creatures swimming around us, attracted by our light, I was in an ecstasy of wonder and delight.

Then days would pass without Captain Nemo putting in an appearance, and none of the crew were ever to be seen. But the *Nautilus* kept on its journey, which, I learned, took us to the Torres Strait, the Papuan coast, through the Red Sea, through a subterranean strait, under the Isthmus of Suez, to the island of Santorin, the Cretan Archipelago, to the South Pole, on whose sterile wastes Captain Nemo reared his black flag with a white 'N' upon it, and through the Gulf Stream.

I cannot here enter into particulars of these wonders of the deep, these amazing and beautiful specimens of unknown life that passed before my vision on this strange journey, and which had never before been seen by the eye of any naturalist. But it must not be supposed, prisoners though we were, that we never emerged from the interior of the *Nautilus*.

One of my first surprises, indeed, was to be invited by Captain Nemo to accompany him on a hunting expedition in the marine forest that grew about the base of the little island of Crespo, in the North Pacific Ocean. We were told to make a hearty breakfast, as the jaunt would be a long one. This we did, for we had soon become accustomed to the strange food, every item of which was produced by the sea.

For our submarine excursion we were furnished with diving-dresses of seamless indiarubber, fitted on the shoulders with a reservoir of stored air, its tubes opening into the great copper helmet. We even had powerful air guns and electric bullets, which proved weapons of deadly precision. When inside our diving-dresses, we could not move our feet on account of the enormous leaden soles, so that we had to be pushed into a compartment at the bottom of the vessel, and the iron doors secured behind us. Water was then pumped in, and we could feel it rising around us, until the compartment was full, when an outer door opened, and we stepped on to the floor of the sea.

For some considerable distance we walked along sands of the most perfect smoothness, and then had to make our way over slimy rocks and treacherous masses of seaweed before we reached the fairy-like forest under the sea, where all the branches of the marvellous growths ascended perpendicularly.

It was indeed a rare experience for me who had written *The Mysteries of the Great Submarine Grounds*, thus to see, at first hand, the life of which I had only been able to speculate on before. We captured many rare specimens, and shot a fine sea-otter, the only known quadruped that inhabits the rocky depths of the Pacific. It was five feet long, and its skin worth a hundred pounds.

So constantly was I enchanted with the wonders of our journey that day succeeded day without my taking note of them; but Captain Nemo, for all his kindness, still remained as mysterious as the Sphinx. One day he became violently agitated after looking through the glass at a point indicated by his lieutenant, and I and my companions were immediately imprisoned in darkness, as we had been when first taken into the *Nautilus*. When I awoke next morning the captain took me to see a wounded Englishman whose head had been shattered, and on my stating that the man could not live for two hours, the dark eyes of the Captain seemed to fill with tears. I thought that night that I heard sounds of a funeral hymn and next day I was taken to a submarine forest of coral, where they buried the man. This was really a little cemetery beneath the sea, as I gathered from the coral cross which had been erected there. Ned Land, unlike me, was soon satisfied with what he had seen of the submarine world and had now only thoughts of escape; but no opportunity had yet offered. We were sailing up the eastern coast of South America, and by 17 May were some five hundred miles from Heart's Content. There I saw, at a depth of more than fifteen hundred fathoms, the great electric cable lying at the bottom of the ocean. The restlessness of poor Ned Land was at its height when he had a glimpse of the American shore; but Captain Nemo bent his course towards Ireland, and then southward, passing within sight of Land's End on 30 May.

Presently the *Nautilus* sank to the bottom of the sea. When at rest the lights were put out and the sliding panels opened. We could now see on our starboard the remains of a sunken vessel, so encrusted with shells that it must have lain there a great many years. As I stood there wondering what might be Captain Nemo's reason for his manoeuvres, he came to my side and, speaking slowly, said:

'That was the *Marseillais*, launched in 1772. It carried seventy-four guns, and fought gallantly against the *Preston*, was in action again at the siege of Granada, and in Chesapeake Bay. Then in 1794 the French Republic changed the vessel's name, and it joined a squadron at Brest to escort a cargo of corn coming from America. The squadron fell in with an English man-o'-war, and seventy-two years ago to this very day, on this very spot, after fighting heroically, until its masts were shot away, its hold full of water, and a third of its crew disabled, this vessel preferred sinking, with its 356 sailors, to surrendering. Nailing its colours to the mast, it sank beneath the waves to the cry of "Long live the Republic!" '

'The *Avenger*?' I exclaimed.

'Yes, the *Avenger*. A good name!' said the captain with a strange seriousness, as he crossed his arms.

I was deeply impressed with his whole bearing while he recalled these facts. It was clearly no common spite against his fellow-men that had shut up Captain Nemo and his crew in the *Nautilus*.

Already we were ascending, fast leaving the grave of the old *Avenger*. When we had reached the surface we could see a vessel steaming towards us. A low boom greeted the *Nautilus* as its upper part showed above the water. Ned Land, aflame once more with hope of escape, made out the vessel to be a two-decker ram, but she showed no flag at her mizzen. It seemed for a moment there might just be some chance of escape for us three prisoners, and Ned declared he would jump into the sea if the man-o'-war came within a mile of us. Just then another gun boomed out. She was firing at us.

It flashed across my mind at that moment that those on board the *Abraham Lincoln*, having seen the effect of Ned Land's harpoon when it struck the *Nautilus*, could not but have concluded their enemy was no monster of the deep – though indeed a monster of man's contriving – the warships of all nations would now be on the look-out for the *Nautilus*, and we on board it could scarcely hope for mercy.

The shot rattled about us as we stood on the opened upper deck of the submarine, and Ned Land, in a mad moment, waved his handkerchief to the enemy, only to be instantly felled by the iron hand of Captain Nemo. Then, frightfully pale, the Captain turned to the approaching man-o'-war, and, in a voice terrible to hear, cried: 'Ah, ship of an accursed nation, you know who I am! I do not need to see your colours to know you. Look and see mine!'

So saying, he unfurled his black flag, and then sternly bade us go below, and the propeller of the *Nautilus* was soon lashing the water into a creamy foam, taking us beyond the range of fire. I held my peace for a time, but after some deliberation, ventured to go up in the hope of dissuading Captain Nemo from more destruction. His vessel was now coursing round the other ship like a wild beast manoeuvring to attack its prey, and I had scarcely spoken when the captain turned on me fiercely, commanding silence.

'Here I am the law and the judge,' he said, almost in a shriek. 'There is the oppressor. Through him I have lost all that I have loved, cherished and venerated – country, wife, children, father and mother. I saw all perish! All that I hate is represented by that ship! Not another word!'

In the face of such fierce hatred it was useless to try persuasion. I and my companions resolved to attempt escape when the *Nautilus* made the attack. At six next morning, being the second day of June, the two vessels were less than a mile and a half apart. Suddenly, as the three of us were preparing to rush on deck and jump overboard, the upper panel closed sharply. Our chance was gone!

Next moment the noise of the water rushing into the reservoir indicated that we were sinking, and in a moment more the machinery throbbed at its greatest speed as the *Nautilus* shot forward under the sea. Then the whole submarine trembled; there was a shock, and then a rending jar above. The terror of the seas had cut its way through the other vessel like a needle through sailcoth! Horror-stricken, I rushed into the saloon and found Captain Nemo, mute and gloomy, standing by the port panel, which had instantly been slid back, watching with a terrible satisfaction the injured vessel sinking with all its crew beneath the waves. The *Nautilus* sank with it, so that its terrible captain might lose nothing of the fascinating horror presented by the spectacle of his victims descending to their ocean grave. When we had seen all, he went to his room and I, following him, saw on the wall the portraits of a woman still young, and two little children. He looked at them, and as he stretched his arms toward them the fierce expression of hate died away from his face. He sank down on his knees, and burst into deep sobs. I felt a strange horror for his man, who, though he might have suffered terribly, had no right to exact so terrible a vengeance.

The *Nautilus* was now making its top speed, and the instruments indicated a northerly direction. Whither was it flying? That night we covered two hundred leagues of the Atlantic. Onward we kept our course, the speed never lessening, and for fifteen or twenty days, during which we prisoners never saw the captain or his lieutenant, this headlong race continued.

Poor Ned Land was in despair, and Conseil and I had to watch him carefully lest he might kill himself. One morning he said to me: 'We are going to fly tonight. I have taken the reckoning, and make out that twenty miles or so to the east is land. I have got a little food and water, and Conseil and I will be near the opening into the small boat at ten. Meet us there. If we do not escape, they shan't take me alive.'

'I will go with you,' I said. 'At least we can die together.'

Wishing to verify the direction of the *Nautilus*, I went to the saloon. We were going N.N.E. with frightful speed at a depth of twenty-five fathoms. I took a last look at all the natural marvels and art treasures collected in this strange museum, a collection doomed to perish in the depths of the ocean with the man who had made it. Back in my own room I donned my sea garments, and placed all my notes carefully about my clothing. My heart was beating so loudly that I feared my agitation might betray me if I met Captain Nemo. I decided it was best to lie down on my bed in the hope of calming my nerves, and thus to pass the time till the hour determined upon for our attempt. Ten o'clock was on the point of striking, when I heard Captain Nemo playing a weird and sad melody, and I was struck with the sudden terror of having to pass through the saloon while he was there. I must make the attempt, and softly I crept to the door of the saloon and softly opened it. Captain Nemo was still playing his subdued melody; but the room was in darkness, and slowly I made my way across it to the library door. I had almost opened this when a sigh from him made me pause.

He had risen from the organ, and as some rays of light were now admitted from the library, I could see him coming toward me with folded arms, gliding like a ghost rather than walking. His breast heaved with sobs, and I heard him murmur these words, the last of his I heard: 'Enough! O God, enough!' Was it remorse escaping thus from the conscience of this mysterious being?

Had I not seen it begin with the tears in his eyes at the death of the Englishman whom he had buried in the coral cemetery, and who was doubtless a victim of one of his acts of destruction?

Now, rendered desperate, I rushed into the library, up the central staircase, and so gained the opening to the boat where my companions were awaiting me. Quickly the panel through which we went was shut and bolted by means of a wrench which Ned Land had secured. The opening of the boat was also quickly fastened after we had got inside, and the harpooner had begun to undo from the inside the screws that still fastened the boat to the *Nautilus*. Suddenly a great noise was heard within the submarine. We thought we had been discovered, and were prepared to die defending ourselves. Ned Land stopped his work for the moment, and the noise grew louder. It was a terrible word, twenty times repeated, that we heard. 'The maelstrom! The maelstrom!' was what they were crying. Was it to this, then, that the *Nautilus* had been driven, by accident or by design, with such headlong speed? We heard a roaring noise, and could feel ourselves whirled in spiral circles. The steel muscles of the submarine were cracking, and at times in the awful churning of the whirlpool it seemed to stand on end. 'We must hold on,' cried Land, 'and we may be saved if we can stick to the *Nautilus*.'

His anxiety was now to make fast the screws that bound the boat to the submarine, but he had scarcely finished speaking when, with a great crash, the bolts gave way, and the boat shot up, released from the larger vessel, into the midst of the whirlpool. My head struck on its iron framework, and with the violent shock I lost all consciousness.

How we escaped from that hideous gulf, where even whales of mighty strength have been tossed and battered to death, none of us will ever know! But I was in a fisherman's hut on the Lofoden Isles when I regained consciousness. My two companions were by my side, safe and sound, and we all shook hands heartily. There we had to wait for the steamer that runs twice a month to Cape North, and in the interval I occupied myself revising this record of our incredible expedition in an element previously considered inaccessible to man, but to which progress will one day open up a way.

I may be believed or not, but I know that I have made a journey of twenty thousand leagues under the sea.

Oh, 'twas in the Broad Atlantic

Anon.

Oh, 'twas in the broad Atlantic,
'Mid the equinoctial gales,
That a young fellow fell overboard
Among the sharks and whales.
And down he went like a streak of light,
So quickly down went he,
Until he came to a mer-ma-id
At the bottom of the deep blue sea.

The Depth of the Sea

Kapitan a D. Ernst Hashagan

Since the days of Jules Verne the submarine has become a formidable reality. It is now a most important and valuable part of the British Navy – and of other navies too, and especially is this so in times of war. A common bond binds men of courage and whichever side they find themselves fighting on they become rivals rather than enemies. Kapitan a D. Ernst Hashagan was an officer in the Imperial German Navy and his job throughout the war had to do with submarines. This is a story he tells of the English Q or Mystery Ships used by the British during World War One. These Q or Mystery Ships were ships that were got up to look like merchant ships. They carried a naval-trained crew and the ship itself was heavily gunned beneath its camouflage. The crew were not uniformed but were fitted out in a rough assortment of clothes in keeping with the appearance of the ship. The drill was for them to set out to entice a German submarine to attack them, having carefully rehearsed how to dodge the torpedo hitting them in too vulnerable a spot. When they had obtained their objective some of the crew, known as the 'Panic Party', rushed for the boats and pulled rapidly away as if abandoning it. It was then presumed that the U-boat would surface, and when it did the ship would cast off its disguise and destroy the U-boat. A number of enemy submarines were destroyed in this way. U-boat Commander Kapitan Hasahagan ran up against one of them and lived to tell the tale.

On 30 April 1917, I was cruising in fine weather on the western approaches of Ireland. A faint patch of smoke hung over the southern horizon; two masts as fine as needle points appeared, and then suddenly a funnel and smoke again. We dived and made our way towards the unknown ship. In half an hour we could already see something more.

A black steamer reeling along like a drunken man, steering a wild, zigzag course. We approached nearer and nearer. She was altering her course every ten minutes, and each time by fifteen degrees. Why? We counted and calculated, puzzled and guessed the reason for her zigzag system and figured out a mean course for ourselves to cross hers before she passed. This flying a common, rather tattered, red ensign was remarkable, I thought. The British don't use that as a rule. In fact, there is a vague something in the air that calls for caution. The coxswain beside me has his hand on the lever of the periscope motor, following my orders.

'Run in the periscope! In: deeper yet! Out a shade. In, in!'

So the eye is always under water, only coming clear of the spray for fractions of a second. The enemy is also altering course again; I see her pause as if trying to make up her mind; then she comes straight towards us. It is as still as death in my boat. Only the rattle of the hydroplanes and the laconic orders for the periscope. And then the vessel's bows swim into the periscope's field.

'First tube, stand by!'

'First tube, shoot!'

There is a slight shiver forward in the boat. The torpedo has gone. It is a dead shot. I feel it. Twenty seconds pass and then come two heavy detonations with violent shocks to us in the boat. We retire to ten fathoms, come to periscope depth again and look round. The ship is heavily buckled amidships. There is no sign of the crew. Then, as I cross her bows with my periscope well out – a decoy manoeuvre – a heavy fire is opened on us from a concealed gun under the bridge. So my suspicions are confirmed, for I also see now another very well-masked heavy gun in the stern. The concealing flaps have been splintered and pushed aside by the explosion of the torpedo. There is no doubt any longer what sort of ship we have disposed of. It is the submarine decoy, Q12.

The end of Q12 was short. The game was up and the crew took to the boats. The panic party did not come properly into action. The stage of the play appeared to be sinking. At a safe distance we surfaced, and I photographed the wreck from three fathoms below the surface. Then we set about it with two powerful guns. It split right in two amidships and went down in separate parts amidst a cloud of smoke and flame. I took its Commander, Commander Lewis, a prisoner, and left his men to be picked up by some passing vessel. But our main task was not, of course, to do battle with these Q boats, but to prowl about and sink merchant ships with valuable cargo.

In August 1917, I worked at the entrance of the St George's Channel. The breeze was strong and the sea heavy. Early in the morning we attacked a steamship. The torpedo track was visible under the target, but no explosion followed. I entered in my log: unexplained misfire. We surfaced, but after a few minutes' fresh air we were attacked by an aeroplane and dived again. It notified its arrival with a load of bombs. Luckily for us there was a few feet of water between us and the bombs, but the force of the explosion upset my boat to such an extent that the bows came right out and the whole conning tower broke surface like the dorsal fin of a shark. But our luck was in. The airman had dropped all his bombs already and had nothing solid enough handy for this lovely target.

Towards dawn we were feeling our way to the much used steamship track between the Kish Lightship and Holyhead. We could not complain of boredom, since the sea around us was alive with destroyers, patrols and steamships. A big, deep-laden ship was our objective. She seemed to be making for Liverpool; but she was escorted by two destroyers and it was obviously a difficult matter to get at her.

In the pitch darkness we had already got too close to one of the destroyers and evidently some of the oil smell from our Diesel exhausts had reached her nostrils. At all events, both destroyers became greatly excited.

Then there suddenly appeared a great light to starboard, astern. My hand flew by instinct to the alarm gong, but hesitated a moment. What was it? Had they seen us? Were we in urgent danger? We watched; then through the night flashed the abrupt Morse signal: 'Keep a sharp lookout. German submarine ahead!' One destroyer is warning the other above our heads! A ray of the enemy's searchlight hovers uncertainly on our bridge. My men laugh. 'They'll never see us, Captain!'

I decided to hold out a few seconds longer. Like a wraith, the slim outline of the second destroyer passes by and fades into the darkness. Behind it, a much larger shape looms out of the night – the steamship! We have broken through the escort and now our way is clear.

Our torpedo takes the ship in the forehold. We stand by,

watching, a few hundred yards away. But what is this? The shadow of the vessel grows larger, grows gigantic. Is she still making headway, then? Like a bull she comes at us, head down. But she is doomed and goes down before striking us.

Until our oil fuel was almost at an end we worked in this way. Then we had to think about making for home.

There were two ways back. Up northwards, round the Orkneys, one door to Germany stood wide open. The patrols might annoy us a little; but we should probably get through all right. To southward there was also a door – the Straits of Dover – but it was locked and barred against submarines by the British barrage, that long, submerged, steel net, suspended from buoys, crossing and recrossing from Dover to Calais. There were mines in front of the net and behind it. There were mines even in the nets themselves – fastened in such a way as to go off if any heavy foreign body bumped in anywhere and excited a pull. There were also sunken blockships for the submarine to ram herself against when she was trying to get through the nets. We didn't know exactly how this barrage system looked beneath the surface; but we had been able to construct a diagram which wasn't, in fact, far from the truth.

The depth of the water between Dover and Calais varied from twelve to twenty fathoms. It only obtained a depth of twenty-six fathoms in one place, and this place was called the deep-water channel. We assumed that the nets would not hang right down to the sea bottom at this point and that a submarine which had failed to slide unseen over the barrage on the surface, through the floats and the buoys, on a dark or misty night, would be able to dive right under it through this deep-water channel. Should we go by the Orkneys or risk the barrage?

There would be more likelihood of doing successful work en route this second way and I reckoned up on my fingers that it would take us two days to get to the western entrance of the Channel, and we could put in four days' work by Cape Lizard or Start Point before we made the run through the barrage.

I glanced at the calendar. Full moon! There it stood, inexorable. So to get over the barrage on the surface might be extremely difficult – the more so if it was calm and clear; but then the weather lately had been stormy and misty, and we stood a good chance of being sheltered by dark clouds.

We decided to take the risk. If we failed to surface the barrage we would dive for the deep-water channel.

By the evening of 31 August we had done our work and arrived at the Colbert Bank Lightship. What was the weather going to be like for getting through? We knew soon enough.

A reddish light shone out behind the French coast which lay clearly visible in the calm night. It was the moon which mounted swiftly, red and gigantic into the sky. Never have I seen such a moon! Its ugly face hung, with a mocking grin, above us and the battlefields of the Western Front.

'Come on, you Germans; come along! I'll light you home all right, so that you'll never forget this night,' it seemed to say.

On the bridge we looked at one another. This 'face' was grotesque, uncanny. It turned night into day. A clear, cloudless sky, quiet sea, sharp horizon and the great monstrous arc-lamp in the heavens. Well, never mind; we may get over the barrage unseen yet with a bit of luck.

'Both engines full speed ahead!'

We proceeded unmolested on the surface as far as Boulogne. The lighthouses of Cape Gris Nez and Dungeness, greatly dimmed, shine across at us. Off Cape d'Alprecht we are confronted by a host of vessels and forced to dive.

Soon after midnight we surface again: all hands remain at the diving stations, and at high speed we make for the barrage, which is about ten miles off. Now or never!

Then, I observe in my powerful night glasses a tiny dot to the northward on the moonlit sea. There is scarcely a second for thought. It is a submarine chaser! In less than half a minute the sea has closed smoothly above us: and only just in time. The enemy shoots up, with a big bow wave.

'Look out! Depth charges coming,' I call down the voice pipe to the control room.

Already now their thunder is tearing the silence to pieces and making a maelstrom of the sea round the boat. All round us roar the bombs. The range is considerable, but we experience a most violent concussion in the boat. So the barrage patrols are already on the *qui vive* and a surface breakthrough is now impossible. We must take our last remaining chance; remain submerged and dive under the whole area of nets and mines, but we are still such a long way from the barrage that we shall have a great deal of difficulty now in finding our gap, the deep-water channel.

From now on the submarine chaser follows faithfully. He is hot on our trail. We can plainly hear his screws above, behind and alongside us. No doubt Very lights are mounting into the sky, searchlights playing, wireless sparking out:

'Hullo! Enemy submarine close in front of barrage steaming to the northward. Look out! Don't let her through. Catch the fish in the net. Bring her up alive or send her down dead, but don't let her through.'

We are all gathered in the control room, with the chart of the barrage before us. The giant hull of our boat is humming along close above the bottom. Every now and then it hits the bottom with a jerk that throws us off our feet. At ninety feet the pointer jumps and trembles. Still ninety feet? Surely it must get deeper soon. Or is the tide setting more strongly on to the French coast tonight, so that we should be keeping out more? We steer down again, always with the feeling that we must 'duck our heads' as low as possible to get under the nets. This time our impact with the bottom is unpleasantly severe. Eighty feet. Good God! it is getting shallower instead of deeper. How can that be possible?

Every sense works feverishly. In our mind's eye we see the net, wide stretched, hanging before us – the gap must be to the left. Port, more to port. At that moment the starboard side of the boat scrapes heavily against some object on the bottom. There is a thunderous hammering as if great blocks of rock are rolling over us, the boat lists over heavily, rights herself, strikes bottom heavily again. Eighty feet still!

We must be close in front of the barrage now.

'Hard to port,' I yell to the helmsman and brace myself instinctively against some fearful power that would impose its will on us. Something quite out of the ordinary must happen now: that alone can save us. My best helmsman, who is always at the helm when danger threatens, quietly turns his wheel as if he were standing in the sunshine. Degree by degree, the boat swings round the compass card. The tide is now pushing us broadside on towards the nets. But no matter. Either we find deep water within a very few minutes, or else we run straight into the deadly arms of the nets and mines.

Again the boat strikes bottom: ninety-five feet! A sigh of relief; a hundred feet, 120 feet; at last, at last! Now we are in more than 130 feet. 'Hard a-starboard, steer 20.' The boat swings back to her old course. At the last minute, it seems, we have found the deep water. Now we are under the barrage. A loud explosion behind us: the stern gives a heave and the starboard propeller suddenly begins to turn irregularly. But otherwise all is still. Once again we hear propeller noises above us. And so we slip away.

Only when hours have gone by and all remains silent, do we come nearer to the surface and take a look round. At daybreak we surface. We are alone and – through!

On 2 September we run safely into Heligoland. One blade of our starboard propeller is heavily dented and split. The starboard after hydroplane is bent upwards. We must have fouled the net with our starboard side and taken a piece of it with us. It was that pull which exploded a charge some distance off, and merely knocked away some of our 'trimmings' without damaging us seriously.

We lay in the dock very well pleased with ourselves. Our 'whale' had had a sharp rap on his tail! But he had only given himself a good shake and come through.

The First Submarine Passage of the Dardanelles

Commander H. G. Stoker, D.S.O., R.N.

As all is fair in love and war, it is only right that we should listen to Commander H. G. Stoker as he recounts his experience of a like hazard: that of passing through the heavily mined and fortified Dardanelles Straits in the same World War. The Straits are thirty-five miles long and half a mile wide at their narrowest point with a current running through them of three to five miles an hour. It was necessary at that time that the Dardanelles should be forced in order to destroy enemy shipping in the Black Sea. Several submarines had tried it and failed to get further than the narrow point. An E-Class submarine could travel only fifty miles underwater, so, with a strong current running against it, this made the operation extremely hazardous, if not impossible. But Commander Stoker, in an AE2-class boat, believed he could do it. He based his confidence on the hope that the current was strongest at the surface and that it would be weaker at depth. There would be mines, too, of course, only to be dodged by deep diving, at which depth the periscope could not work and so they would be 'blind'. The Admiralty did not share Commander Stoker's confidence and at first permission to attempt this feat was refused, but later as the war reached a critical stage the refusal was retracted. 'If you succeed,' he was told, '. . . it may well be that you have done more to finish the war than any other act accomplished.'

AE2 prepared for sea and two hours later they threaded their way through the crowded harbour on their perilous assignment.

It was after midnight when we made our way to the Dardanelles' entrance – a lovely night, clear and calm, with a fair-sized moon and myriad stars casting their ghostly light on the smooth water.

Our plan was to enter the Straits after the moon had set, and proceed slowly along the surface – slowly so that our wash, showing white in the darkness, would not attract attention from the shore; and on the surface, so as to conserve electric power for diving. We hoped in this manner to cover some miles without being discovered by the enemy. Then, at break of day, we would dive.

As we arrived abreast the entrance the moon had still a small distance to go before dipping into her silvery bath.

We waited by the side of the black and sinister-looking destroyers, while the perfect stillness of the night seemed to heighten the tension of the moment. And then the moon touched the horizon, and was gone. In the darkness we crept away from the destroyers. One man only was on the bridge with me, all others below ready for instant diving, as we followed our course in the centre of the Straits at a speed of seven knots.

Two enemy searchlights on the southern shore were sweeping the waters with their long rays; further on the more powerful searchlight at the Narrows threw beams of a yellower hue as it searched the higher reaches. As we drew nearer and yet nearer to the first light it seemed as if progress on the surface must soon be stopped if we were to remain undiscovered.

And then the Goddess of Fortune smiled; the first light spluttered and went out. We crept along with a renewed sense of security. And then almost too unbelievable to be true, the second light quietly faded out. It seemed tremendous luck – we could now, in practical safety, follow our surface route much farther than we had ever dreamed possible, perhaps right up to the main minefield.

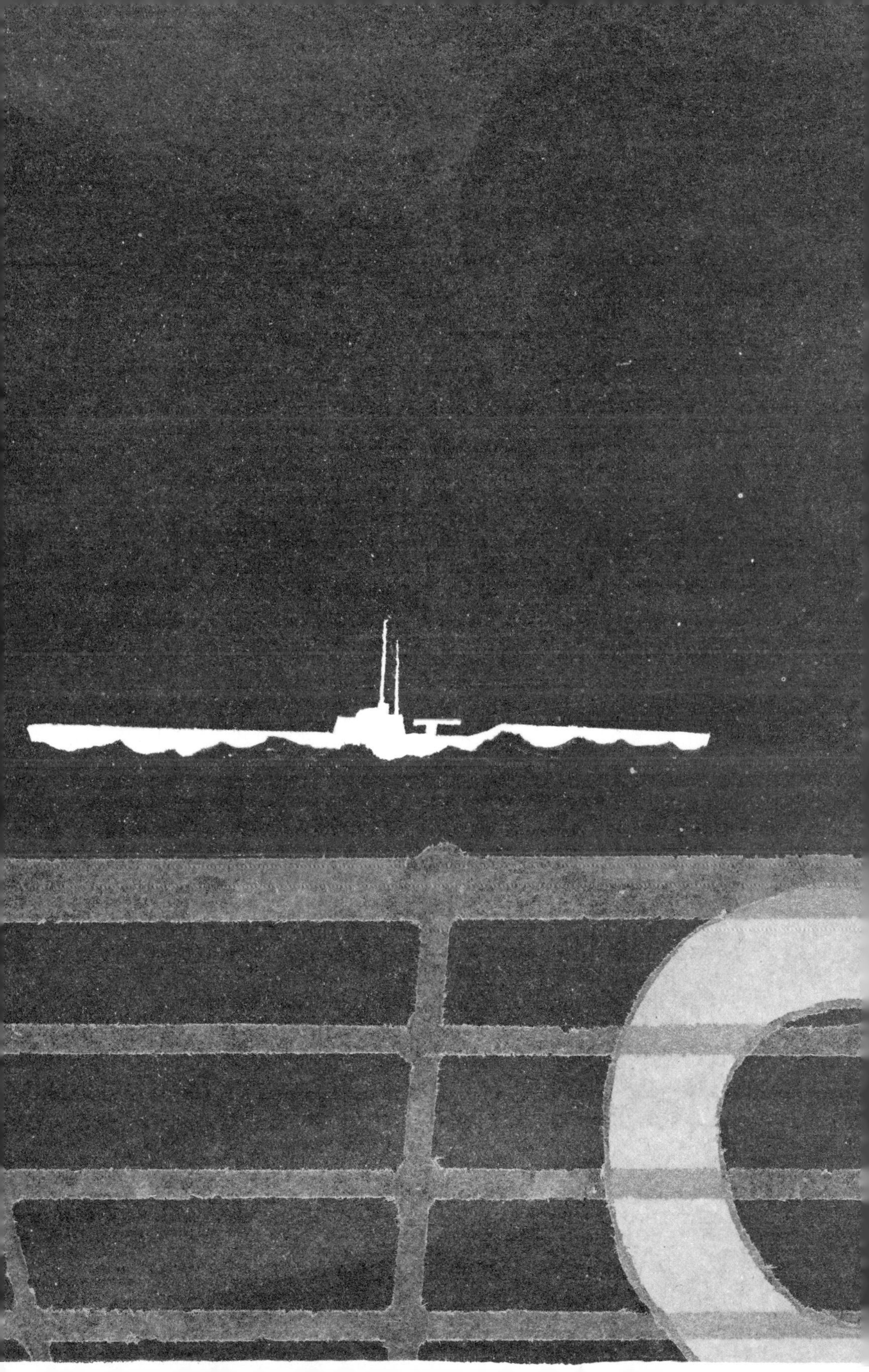

On and on we crept until, of a sudden, the maddeningly clear rays again shot across the water. It would mean instant discovery to remain on the surface, so we must dive. I closed the conning-tower lid, gave the order to flood the tanks, and went below. And then – moment ripe – an accident happened. The shaft which worked the foremost diving rudders broke: impossible to mend under several hours' work, impossible to dive, impossible to go on. We must turn back! After all the working up to that moment; with a job before us that might have an important bearing on the outcome of the war; a chance of a lifetime before us – well, one did not like having to turn back. We raced along at full speed, trying to regain the open sea before day broke and the enemy opened fire.

The cold light of morning showed up plainly on the shore as we passed the entrance and rejoined the watching destroyers. We proceeded to an anchorage and feverishly set to work to make good the damaged shaft. By afternoon it was repaired, and all was ready again. But would we be given a second chance?

Towards evening the battleship *Queen Elizabeth* arrived, and the Admiral sent for me. If I'd had a tail it would most certainly have tripped me up as I climbed on to the flagship's quarter-deck. But kindness again was all I met. 'It was very bad luck' the Admiral said. 'You did well to get so far. Try again tomorrow. If you succeed in getting through, there is nothing you could possibly want that we will not do for you.'

There was, however, to be a minor alteration in the orders. The morrow was the day of disembarkation of our attacking army. Battleships would be in the entrance to the Straits covering the landing, and therefore the enemy would probably launch floating mines against them. So, instead of trying to pass through unseen, AE2 was to attack and sink, if possible, any mine-dropping ships found.

'In fact,' said the Chief of Staff, 'generally run amok at the Narrows – if you get there.'

If you searched the world over I doubt if you would find a more unpleasant spot to carry out a submarine attack than this Narrows at Chanak. Half a mile wide, with the current running at its strongest, it is certainly not an ideal place for manoeuvres in a comparatively slow-moving and slow-turning submarine. Also the thought that we ourselves might meet one of these floating mines hardly added to the entertainment the day was likely to provide for us. However, we were either going to get through or else we were not, and an extra difficulty or so at this stage did not seem to make very much matter. But as it turned out the new order had a far reaching effect on what followed.

Three a.m. on Sunday 25 April 1915. It was absolutely dark, still, and dead calm as AE2 entered the Dardanelles Straits again and crept along on the surface. With broken clouds shutting out such light as a moonless sky gives, the searchlights seemed more powerful than before. One felt forced to edge away from the lights and nearer to the northern shore. Each time a beam of light touched AE2 with a brighter and yet brighter finger, we held our breath. Would the steady sweep pause for a moment, and show a suspicion of our shadowy presence? There was no hope of reaching as high a point as on the previous night. But as we got used to the eerie feeling caused by the passing light, a necessary boldness forced us farther and farther along, now at dead slow speed on one engine.

Suddenly there was a BANG, and the broken swish of a shell as it hurtled past close to my head. We had edged too near the northern shore, and had been sighted by the lookouts of a gun battery. Within a minute we were submerged. Above us the darkness prevented sight through the periscope, but a faint glimmer of light in the eastern sky gave promise of approaching day. At dead slow speed, and at twenty feet, we dived along on our course, until the gathering light showed giant contours of the hills of the northern shore. Then we lowered the periscope and plunged to seventy feet for the passage through the main minefield.

It is difficult to describe our experiences and feelings during the next hour. The rappings and scrapings on the hull of the boat by the mooring wires of the mines, held taut by the buoyancy of the mines themselves overhead, seemed most damnably continuous. One ought to be fairly safe when well submerged, but there was always the chance of one of the wires catching on a projection on the boat's side and dragging its mine, with a bang, down on the top of us.

On two occasions something hard – much harder than the wire – hit the bows, and then went tap, tap, tapping along the boat's side. We imagined, uncomfortably, that they were mines failing to explode. Once some object got caught up forward and remained knocking insistently on the starboard bow. For several minutes we all listened to it in uneasy silence before it broke away, tapped along the side, and followed the rest of our enemies astern.

Twice we rose in the minefield for hasty observation and quick correction of course, then back to the safer depths. The observations showed we were progressing at faster speed than anticipated – the current was evidently much weaker at the seventy feet depth. Even so, I was agreeably surprised on rising the third time to find we were already through the minefield. The Narrows were only three or four hundred yards ahead.

The revised order to attack mine-droppers now obliged us to keep the periscope up for a considerable time to take stock of the situation.

The surface of the water was a flat and oily calm, so the periscope was immediately sighted, and a heavy fire opened from the forts on either side. The shock of projectiles striking the water overhead caused subdued thuds in the submarine, while shrapnel bullets falling through the water on the boat's deck sounded like hailstones. The water around the top of the periscope was lashed with white spray. The effect was curiously pretty, but added little to the ease of taking observations.

An old battleship hulk from which mines might be dropped was anchored abreast Chanak. Higher up the Narrows, approaching at high speed, were a number of destroyers and small craft.

I decided to attack the old battleship, so I lowered the periscope and edged towards her. As soon as we hoisted the periscope again the hail of fire reopened, and a small cruiser came hurrying out from behind the hulk. Now this was just the kind of vessel that would be fitted as a mine-dropper, and from the course she was following it seemed she was trying to drop mines across our bows. Obviously a better quarry; so I got the sights on at a range of three hundred yards and fired the bow torpedo at her. One of the destroyers was now very close, attempting to ram us on the port side, so at the moment of firing I ordered seventy feet. A last glance as the periscope dipped showed the destroyer apparently right on top of us, and then, amidst the noise of her propeller whizzing overhead, came the big explosion as our torpedo found its mark. We had now two dangers to think about; the danger of not being deep enough to avoid the destroyer and the danger of becoming entangled in the sinking ship ahead – as a ship of that size must be expected to sink very rapidly. So we altered course a point to starboard with the object of passing astern of her.

The risk of remaining off one's true course for any length of time in such narrow and fast running waters was obvious, and after three minutes we altered back to what seemed the correct course for regaining the centre of the Straits. At the same time I ordered a rise to twenty feet for another observation.

We had risen to perhaps forty feet when the submarine

struck bottom hard, and slid quickly up to a depth of ten feet. Through the periscope I observed that the position was on the eastern or Asiatic shore very close in, right under the guns of a fort. As I looked, one of the guns fired, its flash coming right round the periscope. It gave the impression that it was firing straight into my eye, so that I involuntarily jumped backwards from the periscope eye-piece. I lowered it quickly and we tried to refloat the boat.

Now, when the depth gauges indicated ten feet there was a considerable amount of the conning-tower and bridge of AE2 above water; indeed, the tops of the periscope pedestals, being the highest objects, were almost ten feet clear of the surface.

With the boat apparently fast aground and a continued din of falling shell, the situation looked as unpleasant as it well could be – very like the end of the story. In reality it was only five minutes before the boat began to move; but even in that time it seemed inconceivable that we were not hit. I afterwards learnt that the guns of the fort could not be depressed enough to bear on us, but the other forts and ships must have made very bad shooting to miss this standing target.

The efforts which eventually proved successful to slide the submarine down the bank left her heading down the Straits again, and we had to turn a half-circle to get her back on her proper course. At a depth of seventy feet we went ahead on the port propeller, helm hard a'port, and dived off the bank trying to turn as quickly as possible. A few minutes passed, while propellers rushing overhead caused pleasant thoughts of the trouble we were making; then we swung rapidly to our proper course, and went ahead on the starboard propeller.

Bump! From a depth of seventy feet we slid gracefully to a miserable eight feet. Where on earth were we now? Through the periscope I observed that AE2, with an apparent liking for forts, had chosen one on the western shore under which to run. The cursed current, which had swept us across to this point, relented for a moment and helped us by swinging the boat's stern round to port, leaving her grounding more aft than forward, and with an inclination down by the bows. A quick glance round showed a gunboat and some destroyers, little more than a hundred yards off, blazing hard with all their broadsides, a cluster of small boats which we guessed were picking up the survivors of the sunk cruiser, and then, best of all, a clear view up the Straits showing that if we could only get off we were heading on the right course.

Full speed ahead both motors! Ominous noises from aft made one fear the propellers would get smashed, but on we must go! – and after a shake, then a move, then another shake, AE2 gave two great bumps and slithered down to thirty feet – having been four minutes at the eight-feet depth. Again the escape must be considered miraculous.

Away, then, at seventy feet, with a host of small vessels in close pursuit. The two severe bumps might have caused leaks in our ballast tanks, and we feared the submarine might not be under sufficient control diving; but all seemed well, and, after a spell, we rose to twenty feet to observe.

Right ahead was Nagara Point – the last of our navigational obstacles involving a nasty right-hand turn. Surrounding us were the pursuing vessels – a gunboat, some destroyers, and a number of tugs and small craft. We could only spare time to fix our position accurately before the destroyers, in attempts to ram, became dangerous. Then away to seventy feet.

Now came the problem of rounding Nagara Point. If we grounded while near the surface for observation we could not hope to escape again; we would also be in more danger near the surface of being caught by swirls and eddies of current. Also there was the obvious danger and difficulty of rising for observation amidst so many pursuing craft. So we decided to attempt the turn at ninety feet without making any observation at all.

Fortune favoured us, and when we rose again Nagara Point – the place, it is said, from which Leander's semi-submarine efforts commenced – was abaft the beam, we were heading into wider reaches, and the enemy ships still hurried and scurried behind us. But even as I looked they sighted our periscope, their guns spoke, and the chase was resumed. The water was so damnably calm that we couldn't take even the shortest spell of observation without the periscope being seen.

We dived to seventy feet and made away up the Straits. This time, with a clear run, we could safely remain below for a longer period, and it was three-quarters of an hour before we rose again, hoping to find the pursuit well shaken off.

But no such luck – the enemy were still in close attendance, so close indeed that we feared that we might have caught an observation net and be now towing a tell-tale buoy above our heads. I couldn't see any such buoy through the periscope, but I saw something else instead. Just ahead, not a hundred yards from us, were two tugs, one on either bow with a heavy wire rope stretching between them. It was right across our track. We immediately dived to eighty feet, and turned off to starboard to consider the situation.

The more we thought, however, the less pleasant it seemed. These tugs were laying a trap of some sort, and no doubt there would be more such traps during the twenty-odd miles which we still had to go before we were clear of the Straits. The longer we remained in their unpleasant company the more chance there was of some ordinary diving accident forcing us to the surface and to instant destruction. If we'd caught an observation net our end was certain in any case, and so it was delaying the inevitable to go on. So we turned at right-angles and ran direct for the Asiatic shore. We knew there was a bank that shoaled slowly there, and so we approached it at dead slow speed and grounded and rested on the bottom at a depth of seventy feet.

The most anxious period of the day followed. Had we

caught an observation net? If so, we were marked, and the end must come soon. And even if we hadn't the enemy had only to carry out intelligent sweeping operations of the few places a submarine could hide on the bottom, and they would have only too good a chance of finding us.

It was about 8.30 a.m. when we grounded. An hour later a ship passed overhead and there was a knock at the boat's side as a sweep hit and jumped over us. We were excessively lucky it didn't catch up!

Soon afterwards other ships passed, and this went on at regular intervals. One of these vessels was obviously a single-engined ship and its solitary screw made an easily recognizable noise. We named him 'Percival'. 'Percival's' repeated passages over us were trying for the nerves, and the fact that we were well out of the track of ships following their ordinary course up and down the Straits proved that 'Percival' and his friends were searching for us.

We decided we must move to another place in the hope that the passage of ships overhead would cease. But the diving control of the boat had been lost. The bumps from the last grounding had evidently so strained her that several of the ballast tanks were leaking. It was most difficult to regain control without coming to the surface.

We made several unsuccessful attempts, and then had to resign ourselves to staying where we were, helpless, until darkness could permit us to rise to the surface and – enemy also permitting – readjust the ballast tanks.

Time moved beastly slowly. 'Percival' and his friends passed and repassed at steady intervals. Attempted jokes as to his reappearance fell mighty flat. The few moments immediately after his passing were the bad ones. If any sweep he was dragging after him caught us, the side of our boat would be blown in on us.

All things come to an end. At 6.45 p.m. – that's to say after about ten hours of this anxiety – 'Percival' passed to pass no more. Two hours later AE2 rose to the surface after being submerged over sixteen hours.

We found a bright moonlight night overhead; but, to our relief, no enemy ships in sight. Our position was about half a mile from the Asiatic shore, in the sweep of the bay that lies above Nagara Point. Marshy swamp land, devoid of habitation, afforded safety from observation from the shore. It was unlikely that ships from the centre of the Straits could see us against the land. The only danger of discovery now lay in being found by a vessel patrolling the coast during the three or four hours necessary for us to remain on the surface for recharging batteries.

Now, too, we could signal the fleet. A dramatic moment, that, as we watched the damp aerial wire throwing purply blue sparks as the longs and shorts of the call sign were flashed. But the answering call never came. Wireless in submarines was very unreliable in those days. It was of the utmost importance that we should get into communication with the admiral and tell him that the most difficult part of our task was accomplished. Whether any other submarines should be risked would depend on the success or failure of our attempt, so he must know as quickly as possible that we had now practically succeeded.

This wireless failure was a very great disappointment. All that we could do was to flash out our signal in the hope that some ship would pick it up. And this we did. (Years afterwards I learnt from Admiral Keyes that our signal was received at a critical moment during a Council of War on board the *Queen Elizabeth*, and the news that a submarine had got through altered the whole tide of the discussion.)

We resumed our passage up the Straits before dawn, proceeding slowly on the surface. When the gathering light showed clear enough for an observer to sight us from the shore we dived once again.

Objects were just beginning to take definite form through

the periscope when I sighted two ships approaching, obviously men-o'-war, one in front of the other; the leader, as far as I could see in the bad light, was the smaller. Both had two funnels. They were not far off, and the persicope, which was making a big white wash, must be lowered, for the water was still absolutely calm, unmarked by a ripple.

Steering a parallel and opposite course to the enemy, we approached, and when judgment estimated us to be within torpedo range, hoisted the periscope. Right abeam was a ship, looking mighty big at a range of five hundred yards, and I jumped to the conclusion that she was the second or rearmost ship. The bearing for firing the port torpedo was on, and we fired. The ship dodged, the torpedo passed ahead of her; and then, looking round, I found to my disgust, that I had fired at the smaller of the ships. The other was following, but it was now too late to bring any of the other tubes to bear with a good chance of the torpedo hitting. We had lost a glorious chance, and through my fault alone. Of little use to think that two sleepless nights and the experiences of the previous day hardly tended to produce the even, balanced mind necessary to successful submarine attacks in these unsuitable conditions of bad light and smooth sea. We had had a glorious chance, and it was gone. Terribly disappointed we dived along on our course, forming a resolve to find a quiet spot for rest before carrying out another attack.

Towards seven o'clock we approached Gallipoli town, at the head of the Straits. Stretching in endless chain from shore to shore across our course was a quantity of small boats. Plunging to seventy feet we passed peaceably beneath them, and so dived out into the Sea of Marmora.

Our wish was realized. The submarine passage of the Dardanelles was not only proved possible but made.

AE2 had other hazardous adventures. In the end the enemy guns got her. She now rests at the bottom of the Sea of Marmora; and within her lies a case of most excellent port – which we had never opened.

First Dip

Harry Grossett

Harry Grossett had intended to be a shipwright, a ship builder, but diving was in his blood and after seven years apprenticeship to that trade he knew that diving was all he wanted to do. Nowadays people take to the underwater world for the sheer pleasure and fun of it, of discovering the wonders of this new element: fishing with spear rather than rod and line in a garden wonderland. Others, like Jacques-Yves Cousteau, use it scientifically, for the serious study of marine life. A new and exciting activity is underwater archaeology, and a perhaps more romantic aspect of that is the search for ancient and lost ships. But Harry Grossett's interest was more basic. Simply, he wanted to work under the sea – at any job that presented itself. He was in it for the money, too, and for the chance to travel which went with it. But for whatever reason human beings take to this element, all, sooner or later, have to make their first dive. Here is Harry Grossett's account of the training that led up to his first dive, and, finally, the dive itself.

You have got to be thin if you want to be a deep-sea diver. There is no place for fat men at the bottom of the sea.

This is not a union rule. (There is no deep-sea divers' union – there are not enough of us to form one, and we are too widely scattered.) Thinness is necessary simply because of the great pressure under which a diver has to work when he goes down deep. Only a man of spare build can stand up to it without suffering physical harm.

There are other qualifications for diving, but most of these can be acquired. You need to be born thin – and you have to stay thin; not as thin as a rake, but without any superfluous fat. A deep sea diver has to watch his figure like a ballet-dancer. That is about all the two jobs have in common.

I was lucky enough to be born thin, and I have kept my figure without much trouble. I have never had to do any slimming. I have known other divers who have had to retire because they put on weight. Too much beer is a common cause.

A diver has got to be fit, too. I found out just how fit when I had that first medical examination. I was weighed and measured, tapped and thumped, and cross-examined about my medical history. The doctor tested my heart, my lungs, my eyes, toes, fingers, reflexes – and especially my ears. He looked down them for a long time. Then he asked me about my nerves, and I told him they were fine, but he did not take my word for it. He had an accomplice slamming doors behind me to see if I jumped.

When it was all over – as I thought – I had to go into the gymnasium. Here the doctor had me jumping off benches on to one leg, jumping back again, then the other leg, hopping, skipping, and running round, pulling up on the horizontal bar – and every now and then I had to stop while he took my pulse and sounded my heart. After about half an hour in the gymnasium I was told I was fit, and released.

Nutty Head (my friend) passed his medical, too, and we went together to the naval diving school at Sheerness, H.M.S. *Wildfire*. There were ten naval ratings there – two engine-room artificers, four shipwrights, two blacksmiths, a seaman petty officer, and a seaman gunner. Nutty and I were the only civilians. Courses were continuous, and the pupils were at different stages of training. We had simply come in and filled two vacancies caused by men passing out.

We were still paid by the dockyard, but we were put under the same strict naval discipline as the others. There was no rate or rank in the diving class. We were all called pupils and had the same duties.

A gunnery officer, who was also a qualified instructor diver, was in charge of the class. He had two other instructors

– a chief shipwright and a chief bosun's mate. There was also a medical officer on the staff, and we all had a thorough medical examination once a week.

First of all we learnt the theory of diving. As the theory I was taught in 1903 is almost exactly the same as the theory that is taught to men learning to dive today, I am going to say a little about what I learnt. The first thing they taught me at Sheerness was that we are all living at the bottom of an ocean – an ocean of air. It is several miles deep, and it gets thinner as you go up.

Now air has weight, and weight means pressure; and the air at sea-level is pressed down by the weight of all the air above it. If you go up a mountain, or in an aeroplane, the pressure gets less because there is less air above.

The pressure of the atmosphere at sea-level is not always the same. It is higher in good weather than in bad weather, as you can see from a barometer. When it is fine the glass goes up; when there are storms the glass falls. But the changes are very slight, and generally speaking atmospheric pressure at sea-level is about 14·7 pounds per square inch. At the bottom of a coalmine it is a little more. At the top of Mount Everest, which is nearly 30,000 feet up, the pressure is only 4·6 pounds per square inch.

Most of us take atmospheric pressure for granted, except when we tap a barometer. It does not affect us physically, although our bodies are hollow above the waist. This is because there is always air in our bodies, and it has the same pressure as the air outside. So while there is a weight of 14·7 pounds pressing in on every square inch of our chests, just the same weight is pressing out from inside. If you could take the air out of your lungs with a vacuum pump, your chest would cave in under a weight of about two tons. So long as the air is inside, the pressure is equalized and you do not feel it. Man was made to live in the atmosphere.

But man was not made to go down to the bottom of the sea.

That is the place for fishes that have not got hollow bodies and do not need air. Deep-sea diving is unnatural. For a man to go down fifteen fathoms, taking his air with him, is cheating nature; and nature is never easy to cheat.

Water, like air, has weight; and the ocean of water is like the ocean of air. The deeper you go, the more water there is weighing down from above, and the greater the pressure becomes. On the surface of the sea, where we go swimming, it is not much more than atmospheric pressure. But water is a lot heavier than air, and the pressure increases much more quickly with depth. If you could go down a coalmine 2,200 feet deep, the pressure of the air at the bottom would be about sixteen pounds. You only need to go down three feet of water to reach that pressure. Every foot you go down brings an increase of nearly half a pound per square inch in pressure. At thirty-three feet the pressure is nearly thirty pounds – just double normal atmospheric pressure. And a deep-sea diver has to go down thirty fathoms and beyond.

If you could take an empty tin – empty, that is, except for air – seal it hermetically, and send it down thirty fathoms, the pressure of the water would smash it long before it reached the bottom. The same thing would happen to a diver if he went down breathing air at normal atmospheric pressure. His hollow body would be crushed. His diving dress would be no protection, for it is only made of indiarubber. So he has to be 'fed' with compressed air, to equalize the pressure inside and outside his body. The pressure of the air he breathes must always be equal to the pressure of the water surrounding him. That means that the deeper he goes, the more compressed the air must be.

Air is a combination of gases, and gases are generally easy to compress; and breathing compressed air does not have much effect on the working of the body. I have been breathing it for nearly fifty years, on and off, and it has never done me any harm. But it is not a natural thing to do, and nature is a dangerous enemy who never gives in. Even when victory seems easy, you must be constantly on your guard, or she will take her revenge in an unexpected way.

That was what they taught us first at Sheerness. We had our theory lessons in the cellars of the dockyard, which were cold and damp. Perhaps that was appropriate for men learning to be divers. It is cold and damp at the bottom of the sea.

Our diving equipment was treated more tenderly than we were. The diving dresses were hung up on lines in a drying room. At Sheerness I was taught how to look after all the diver's equipment.

Basically this does not differ from the equipment I use today. The only thing new to me at the diving school was the telephone breast-rope, or life-line. The breast-rope was attached to a terminal at the back of the helmet, and then brought up under the right arm in the same way. Telephone wires were embedded in the breast-rope.

The receiver was fixed in the crown of the helmet, and the transmitter was on the side of the helmet between the front glass and the right-hand window. You had to move your head to the right to speak. To 'ring up' you pressed a button with your chin, and a bell, or buzzer sounded in the diving boat. There were many disadvantages about the diver's telephone then, and although it has been greatly improved there are still some now. Every man who learns diving today has to learn the code of signals on the breast-rope and air-pipe.

At Sheerness we were taught to use the telephone, but we made all our signals by hand. I had the advantage of knowing the navy code but had not realized how thoroughly it had to be learnt. We had to keep practising the signals until we could almost do them in our sleep. We had to know them so well that in a crisis our reactions would be immediate and automatic, like the jerk of a foot when the doctor taps you below the knee. Timing is the only difference between two pulls and two bells; but two pulls on the breast-rope means 'Send me a slate', while two bells repeated several times quickly means that the diver is foul and needs the assistance of another diver. When you are in trouble you have little time to think before signalling, and unless you have been trained intensively fear might make you pull the rope madly and meaninglessly.

I have never met a diver who has not known fear, but everyone learning to dive must learn to conquer it. The theme of all our instruction was never to give way to fear. 'Keep cool, calm and collected on the bottom' – that was hammered into us day after day. 'Don't panic. You may feel alone but you aren't. There's a man at the other end of the lines.'

Our instructors told us frankly of the dangers of diving. We were taught how to deal with emergencies like attacks by sharks and octopuses. These seemed rather remote from the cellars of Sheerness dockyard, but I was relieved to learn that they were not so dangerous as the story-books made out. The real enemy, we learnt, was pressure: it was always there.

We learnt the job of attendant as well as diver. We had to act as attendant to each other when we went out together in the diving boat. This was a sailing pinnace, and we had to pull the 15-foot ash oars ourselves. The instructors made us take her out in rough water to give us a chance to find our sea-legs. They also wanted to know if we had any. Luckily I was not seasick. Some were, and got over it. If they had not they would have been sent away from Sheerness. Salvage work often has to be done in a choppy sea, and a seasick diver would be of as much use as a ship without a rudder.

We had medical lectures, and I learnt why the doctor had been so interested in my ears. A diver has got to keep his tubes clear. I am talking about the eustachian tubes – the passages that lead from inside the eardrums to the back of the nose.

Unfortunately most people's tubes are hardly ever clear,

and they are soon blocked by a cold in the head. It is not safe for a diver with a cold to go down very far. He will probably burst his eardrums.

After I had been out in the diving boat two or three times, acting as attendant and learning how to go down and come up, and control the air, I was given my first dive.

I suppose that every diver remembers his first dive. I have forgotten much of what happened at Sheerness, but that memory is still vivid.

I was eager to go down. I was impatient while the attendant helped me to struggle into my diving dress, and still quite happy when the helmet was put on. The pump was hove round a turn or two, and for the first time I heard the sound of the air coming in, pf-pf, pf-pf. I was still impatient when I stood on the ladder and the weights were put on and secured. I told the attendant I was all right, and he screwed in the front glass. Then, suddenly, I felt alone in the world.

I felt myself getting buoyant as the air came in and puffed out my diving dress like a frog's skin. Then the attendant gave me a pat on the top of my helmet, and I went backward down the ladder. As my body went into the water the diving dress was pressed close to me, skin-tight, and the air inside rose and lifted my weights. I felt a sudden lightness.

I let go of the ladder and raised my hands, and then I was drawn by my breast-rope and air-pipe to the shot-rope, the guideline from the boat to the sea-bed. With my head just under the water I closed my outlet valve for a moment, so that the attendant could see if there were any leaks. Then I opened it again, put my hand up in the air, and started to slide down the shot-rope, holding it with both hands and keeping it between my legs.

Probably the water was cold. It was November, which is a poor time of the year to begin diving. I had plenty of woollens under my diving dress, of course, but my hands should have felt cold. I do not think I noticed them. I hardly felt my body at all. I may have swallowed once or twice to make my ears 'pop', but I was not conscious of it. My mind was filled with this growing feeling of loneliness and isolation.

The feeling became stronger as the light went. This happened very quickly, for in Sheerness dockyard, where I had my first dive, underwater visibility was almost nil. At the bottom it was nil, and when my boots landed on the mud I was sure that I was alone in the world. I could see nothing and hear nothing except the chattering of the outlet valve as the air escaped.

I was not alone, of course. I signalled my arrival with one pull on the breast-rope, and felt a bit better when I felt the reply from the attendant above. Then I adjusted my outlet valve.

It is with the outlet valve that the diver controls his buoyancy. He has no control over the air coming in. If he wants more air or less air he has to signal for it. As he goes down the supply of air is increased, for the deeper he goes, the more compressed the air has to be to balance the increasing water pressure outside. When he is on the bottom the air is pumped in at a uniform rate.

By adjusting the outlet valve the diver controls the speed of his descent. When he reaches the bottom he adjusts it again, to give himself enough buoyancy to take the weight off his shoulders, but not too much to lift his feet off the ground. That was what I did; but having reached the bottom, and felt something solid underfoot at last, I was not very generous and kept my outlet valve fairly open.

When I had done this I felt my hand grasped by the instructor who had been waiting for me. We touched helmets, and he asked me if I felt all right.

It is the easiest thing in the world for two divers to talk to each other when they are close to each other under water. All they need to do is to touch helmets, and they can converse in their normal voices. The only thing necessary is for each

to put his finger on the spigot of his outlet valve, to stop the noise of the air escaping from his helmet. Of course the spigot cannot be held for more than a few moments at a time, otherwise the diver's dress would get too full of air and he would start to go up again. The outlet valve is his control over his buoyancy. By opening and closing it he can go up or down at will.

So I pressed the spigot and told the instructor I felt all right, and he asked me if I would like to take a little walk. I could not see him or anything else, but I had the feel of his hand as we moved along on the bottom.

A diver walks in a sort of slow-motion goose-step. You can get some idea of the feeling by walking about in a swimming pool. The difference is that the diver's feet do not come up so easily. He has got a lot of lead on the soles of his boots.

I had made it hard for myself being so mean with the air. I felt compressed – as I was – the weights were heavy on my shoulders, and walking was hard work in that mud. When there is any visibility at all at the bottom the diver has to try to keep as buoyant as possible, for even a slight scuffling with his boots can stir up the mud into a thick fog. This did not matter in Sheerness dockyard, where you could see nothing to begin with. It was like being in a sea of black ink.

I had my little walk, and after a bit the instructor let go of my hand. I was utterly lost, without any idea of direction. But he kept close by me, and eventually led me back to the shot-rope and told me I could go up. I had been down about ten minutes.

We shook hands again, and I gave four pulls on the breast-rope, meaning 'I am coming up'. I went up under my own steam, or rather air. Normally a diver does not need either to be pulled up or to climb up the shot-rope. He simply closes his outlet valve a little, inflating his divingdress and so making himself more buoyant. He lets the air take him up.

He has to watch it, though, for as he rises the water pressure decreases and so the air in his dress expands. I did not have to worry much about this when I came up from my first dive, for I had only been down three fathoms, or eighteen feet. It was not far but to me it had been a journey to another world.

There was no shining sun to greet me, but I liked the look of that dull November sky. When I reached the surface, and was drawn back to the ladder, I felt helpless for a moment, like a fish out of water. Then the weights were taken off, and when my face glass was removed I was able to grin encouragingly at Nutty Head, who was going down next.

Trapped

David Masters

Forgotten or not, first dives are soon left behind in the more serious everyday underwater work. Salvage men are a rare breed: courageous, infinitely patient, with a quiet determination to carry on when other men might give up. They are prepared at all times to take whatever job comes to them irrespective of its difficulties and often its dangers. And not always do they dive for gain. 'Sea-gold', as Philippe Diole puts it, may tempt them, but they will work even more dedicatedly in the cause of their fellow men. The following story is of a dive by brave men to rescue other brave men when the odds weighed heavily against them all.

Although we live in an enlightened age, superstition is still rife, and not many people would care to dive for the first time in a submarine bearing the unlucky number thirteen. Yet in spite of the fact that sailors are generally credited with being more superstitious than most people, no thought of danger crossed the minds of the seventy-three men who during the First World War stepped aboard the British submarine K13 in order to carry out her trials. She was a wonderful craft, 334 feet long, just under 27 feet wide amidships, and as she lay at her moorings she displaced 1,880 tons.

Like her sister ships of the same type, she was one of the fastest submarines afloat, capable on the surface of overtaking most battleships in order to send them to their doom, able to take her place with the Grand Fleet and steam along with them at top speed without being left behind. This wonderful speed was attained by fitting her with steam turbines in addition to the usual oil engines and electric motors. Her stumpy funnels folded down when she was diving, and the introduction of steam made it essential to get fairly big ventilators. In order to dive she could take into her ballast tanks 800 tons of water in four minutes, but with a big submarine over a hundred yards long, all divided into many compartments, diving was a delicate operation that depended for its safety upon all men carrying out their duties instantly. It was necessary that the crew should be quite conversant with their craft and that there should be perfect team work. But an absolutely new craft is bound to present some strange features to her first crew. In this case she was a new development in submarine practice, and it was probably the fact that the K13 was unfamiliar that brought about the ensuing disaster.

Built on the Clyde, she was taken along to the Gareloch to be put through her paces. The Gareloch was quiet, away from spying eyes, free of the attentions of the unwelcome enemy submarine, and here the K13 carried out her surface trials satisfactorily. The conning tower was closed, the funnels were dropped back flush with the deck, and orders were given to trim the boat for diving. The watertight doors were shut and the sea began to flow into the tanks. Then, as the craft submerged, came disaster. A mighty rush of water swept into the after part of the ship, drowning instantly the thirty-one men on duty there, and carrying the K13 stern downwards to the bottom. It was afterwards discovered that in diving some of the ventilating scuttles had been left open and these had flooded the stern of the ship. It was a tragic oversight that in a moment swept thirty-one men into eternity.

In the forward part of the K13 forty-two men were imprisoned, held fast on the sea-bed by the weight of water in the ship. There was no trace of panic. Nobody turned a hair. As quietly as though they still floated serenely on the surface, they stood by and carried out their commander's orders.

For hours they strove to get the ship to move to lighten the tanks sufficiently to bring her to the surface again. The ship remained fast. No trace of movement was to be detected. The watertight bulkhead across the centre of the vessel held death at bay for the moment, but no one knew how long it could withstand the terrific pressure. At the other side of the bulkhead lay their dead companions, and the hungry sea was waiting to engulf the living. Death threatened them from all quarters, death from drowning, death from asphyxiation owing to the exhaustion of their air supply, death from starvation even if the air held out. Hour by hour death came nearer. The realized it only too well, but still they remained cheerful.

When it was seen that all their efforts were useless, Commander Godfrey Herbert, D.S.O., who was in command, and Commander F. H. M. Goodhart, D.S.O., who was aboard to watch the behaviour of the vessel before taking over the command of the K14, conferred and agreed to try to get to the surface, ninety feet above their heads, in order to obtain help. They knew perfectly well that they were probably going to their deaths, that the odds were so tremendously against them that they were not worth considering. They did not think of themselves; they thought only of the forty men caught in that death-trap.

The one way of getting to the surface was through the conning tower. But the terrific weight of the water above closed the lid so tightly that the strongest giant in the world would never lift it. To raise it was beyond the mere strength of human beings. The only way of accomplishing the feat was to let into the conning tower compressed air until the pressure of the air equalled the pressure of the sea, and as the air burst away upwards the gallant officers hoped to be carried to the surface with it.

Quietly they entered the conning tower, and partially flooded it. The compressed air was turned on. Minute by minute the pressure increased, minute by minute the officers waited, wondering if death or life was to be theirs, whether their attempt was to succeed or fail.

So great grew the pressure that the air could no longer be kept within bounds. With incredible strength it burst upwards and Commander Goodhart was dashed violently against the steel sides of the conning tower and killed instantly.

By the greatest good fortune Commander Herbert missed the full force of that deadly upthrust of air. Still, he too, was hurled upwards, and, as the water rushed in and the air gushed out, was carried clean through the conning tower to the surface.

Already the disappearance of K13 was arousing anxiety up above, and a salvage craft had been called to the spot. A couple of men in a boat, noticing the figure of Commander Herbert as he came up in the Gareloch, pulled quickly towards him and dragged him over the side. He was almost dead with exhaustion, and the wonder is that he survived that terrible ordeal.

As soon as he was sufficiently recovered, he gave an account of what had happened and told how the men were trapped in the submarine. The urgency of the case was obvious. It needed no stressing.

Then began a most thrilling salvage fight. It was a fight, not for treasure, but for human life. It was a race against time, a long tussle with death.

Divers dropped down the shot-ropes to the bed of the Gareloch and began to search for the sunken submarine. The light was none too good, owing to the water being fogged with mud, but they were searching only for a short time when the dark hull of the submarine loomed in front of them. They hurried up to it. One drew an axe from his belt, and hammered hard at the side.

Answering knocks came from within, and those waiting anxiously on the surface heaved a sigh of relief as the divers telephoned up: 'We've found her. They're still alive!'

Surveying the wreck, the divers discovered that the bow of the submarine was about twenty feet higher than the stern, which was already covered by a dozen feet of mud. Wading in slime sometimes up to the armpits, the divers worked their way round her, then quickly sped to the surface and reported her position.

At once the experts summed up the situation. The K13 with her stern full of water, covered up aft by a dozen feet of mud, was too heavy to raise bodily. She was well over 3,000 tons, and up to that time nothing like this weight had ever been lifted from the seabed. The only thing to be done, the sole hope of saving the imprisoned men, was to strive to lift the nose of the craft to the surface, while leaving the stern resting on the bottom. Nothing else was possible.

'The first thing to do is to get through supplies of food and air to them,' the salvage officer remarked.

The divers slid down to the bottom and, disregarding all thought of their own safety, laboured hard and long to connect up with the entombed men. They must have broken the endurance record, for one worked for over twelve hours continuously on the sea-bed without taking food, without resting. Time was too precious for them to waste a second.

They realized the risk, but they accepted it as gladly as Commander Goodhart ran the risk which led to his death. They worked until they were ill and dizzy, floundering in the mud, wrestling with giant steel cables.

Forty men were depending on them for their lives. The thought nerved the divers to do prodigious things. It was essential to communicate with the imprisoned men, to let them know that everything possible was being done for them, to strive to sustain their spirits. Commander Kay of the Salvage Section found the way. Sending down a submarine flash-lamp, he instructed the divers to rig it up in front of the periscope. By peering into this instrument the prisoners were thus able to read the messages that were flashed to them in Morse Code, and were made to understand that they were not entirely cut off from the world after all. With many a struggle, the divers managed to open a valve in the hull and to attach a pipe through which food such as Bovril, bottles of hot soup and chocolate, as well as life-giving air, were passed from the surface. All this entailed long hours of endeavour.

The coolness of the men in the submarine was almost unbelievable.

'Send us down a pack of cards to while away the time,' one shouted up the pipe.

The cards were procured and sent down, and these British seamen played cards while Death peeped over their shoulders.

Up until then the men had been carefully conserving their supplies of compressed air, not knowing how long they would need them to keep alive. Now that air was being pumped from the surface, they were able to use what was left of their own supplies to blow all the oil out of the forward tanks. This lightened their craft considerably.

After a terrific struggle, the divers managed to fix mighty steel cables under the nose of the submarine. Salvage craft and lifting vessels strained away. For a time they made no impression. Then slowly the grip of the mud began to relax and the bow of the submarine, lightened by the blowing out of the oil tanks, began to rise nearer and nearer the surface until, about midnight, it broke clear into view.

It was a weird sight. Great arc lamps lit the scene, and under their glare the salvage men attacked the steel hull of the K13 with oxy-acetylene blow-pipes. Everyone was desperately anxious, afraid that the submarine might slip. Under the intense heat of the blow-pipes, the steel grew soft and melted. Gradually, laboriously, the salvors burned their way through the stout outer plates.

They now made an onslaught on the inner hull, directing the flame on the steel shell. The metal glowed and flowed. A rush of air leaped upwards from the interior of the vessel and blew out the roaring flame of the blow-pipe.

'Get us some matches!' the divers called to those above.

Under their very noses a hand from inside the ship suddenly slid through the hole in the metal, the fingers holding up a box of matches.

'Here you are,' said a cheery voice, and the divers knew that all was well.

Another period of strenuous endeavour and the hole in the metal was big enough for a man to squeeze through. Then, as the forty prisoners were helped and carried to freedom, the cheers of the salvage men echoed to the shore.

Never will men be nearer death than those saved from the K13. For fifty-seven hours they were imprisoned in the sunken submarine at the bottom of the sea, for two and a half days they lived with death at their elbows, not knowing when the end would come.

The Drowning of a Town

Dorothy Clewes

With skill, and patience, and good fortune, men and ships can sometimes be salvaged from the sea. But what about a town that is swallowed outright? Not far from where I live, close to the border of Sussex with Kent, is the town of Winchelsea. Perhaps 'town' is too formal a name for so small a community and yet it has the status of a Borough with its own mayor and corporation and privileges conferred on it by royal charter. Ever at the mercy of marauding enemy forces: the French attacking and being in turn attacked, and forest-locked on its northern side so that this was the last part of England to be reached by Christianity. It was from these forests that the great timbers came that built the ships that went out to do battle and to trade, and, as was common in those days, to waylay other traders on the high seas.

In 1066 William the Conqueror landed near here – at Pevensey – and defeated King Harold at Battle, which was so named to commemorate this event. In 1067, after a visit to Normandy, William came back to England, this time landing at Winchelsea. In 1138 Henry II also landed here, while the French landed on many occasions, mostly to plunder and put the town to the torch. But in 1250 a force stronger than man struck.

It was written by a chronicler of that time – '. . . and on the first day of October, the moon upon her change, appearing exceeding red and swelled, began to show tokens of the great tempest of wind that followed, which was so huge and mightie, both by land and by sea, that the like had not been lightlie known and seldom, or rather never heard of by men then alive. The sea, forced contrarie to his natural course, flowed twice without ebbing, yielding such a roaring that the same was heard (not without great wonder) a farre distance from the shore. Moreover, the same sea appeared in the darke of the night to burne as it had been on fire, and the waves to strive and fight together after a marvellous sort, so that the mariners could not devise how to save their ships where they laie at anchor, by no cunning or shift which they could devise. At Hertburne, three tall ships perished without recoverie, besides other smaller vessels. At Winchelsea, besides other hurte that was done, in bridges, milles, breakes, and banks, there were three hundred houses and some churches drowned, with the rising of the water course.'

But the sea had not yet finished with Winchelsea. The stalwart people of the town worked to save what was left and to put it to rights as best they could, and life went on. In 1262, Henry III landed there on his way to the Battle of Lewes, and later Eleanor, wife of Henry's conqueror, de Montfort, was there. 1266 brought Prince Edward to the town to chastise it for piracy – but Winchelsea, along with other coastal towns, made great capital out of this rewarding form of livelihood and I do not think they would have taken over-much notice of the royal reprimand. Then, on 4 February 1287, a great storm arose and the sea completed what it had begun thirty-eight years before. It tore into the town and drowned it, and all the flat land between Pett and Hythe was inundated, and Winchelsea was as if it had never been. Where fishermen had once dropped their nets, shepherds now watch their sheep; where the marauding French overran the land with sword and torch, tourists now with guide books and camera walk over the marshes to climb the steep hill to the new town of Winchelsea which the survivors of that terrible storm built on the pleateau above. It looks over the lost town to the sea which was once its benefactor and its destroyer.

A Cinque Port

John Davidson

John Davidson was one who reflected on this tragic and appalling catastrophe. He must have been drawn many times to Winchelsea, as I have been, by the strange, eerie atmosphere that pervades even now this new town. An air of unreality seems to hover in the air. He must have paused to wonder about that other old town – the people who lived there, and the lives they lived, and looking down from the safe height of the new town, over the green, sheep-cropped marshes, found these thoughts forming in his poet's mind.

Below the down the stranded town
What may betide forlornly waits,
With memories of smoky skies,
When Gallic navies crossed the straits;
When waves with fire and blood grew bright,
And cannon thundered through the night.

With swinging stride and rhythmic tide
Bore to the harbour barque and sloop;
Across the bar the ship of war,
In castled stern and lanterned poop,
Came up with conquests on her lee,
The stately mistress of the sea.

Where argosies have wooed the breeze,
The simple sheep are feeding now;
And near and far across the bar
The Ploughman whistles at the plough;
Where once the long waves washed the shore,
Larks from their lowly lodgings soar.

Below the down the stranded town
Hears far away the rollers beat;
About the wall the sea birds call;
The salt wind murmurs through the street;
Forlorn the sea's forsaken bride
Awaits the end that shall betide.

The Gold Wreck

Charles Dickens

The discovery of an old wreck is thrilling and exciting. Time has given it an aura of mystery and nostalgia for days past and unknown. A new wreck – new in 1859 – has a different impact on us. There is the agony of having been near and yet too far away to have been any help for those in such dire peril. The wreck of the *Royal Charter* so affected a young reporter that the sight of the broken ship and all it conjured up in his mind could only be exorcised by writing it out of his mind at a later date.

On October 25 1859 winds of over a hundred miles an hour screamed across the West Country and up the coast towards Liverpool. Homeward bound to Liverpool from Australia was the *Royal Charter*. She was a 2,719-tonner with iron sides and not the sort of ship to be disturbed by any ordinary storm at sea. But this storm was far from ordinary. In Liverpool Bay, just off the coast of Anglesey, the storm hit the *Royal Charter*. She could make no headway against the storm in spite of her auxiliary engine. Her captain ordered away both anchors. They bit, held, and then suddenly gave way. In she drove towards Moelfre Bay and finally hit the rock reef stern first.

Besides over four hundred passengers and crew, down with the ship went over £300,000 in gold bullion from the Australian goldfields and who can say how many small private hoards of gold from the miners among the passengers. The disaster made front-page news, and one of the journalists who came to the little fishing village of Moelfre was Charles Dickens. He wrote his impressions of the scene in the early pages of *The Uncommercial Traveller*.

Never had I seen a year going out, or going on, under quieter circumstances. Eighteen hundred and fifty nine had but another day to live, and truly its end was Peace on that sea-shore that morning.

So settled and orderly was everything seaward, in the bright light of the sun and under the transparent shadows of the clouds, that it was hard to imagine the bay otherwise, for years past or to come, than it was that very day. The tug-steamer lying a little off the shore, the lighter lying still nearer to the shore, the boat alongside the lighter, the regularly-turning windlass aboard the lighter, the methodical figures at work, all regularly and slowly heaving up and down with the breathing of the sea, all seemed as much a part of the natures of the place as the tide itself. The tide was on the flow, and had been for some two hours and a half; there was a slight obstruction in the sea within a few yards of my feet: as if the stump of a tree, with earth enough about it to keep it from lying horizontally on the water, had slipped a little from the land – and as I stood upon the beach and observed it dimpling the light swell that was coming in, I cast a stone over it.

So orderly, so quiet, so regular – the rising and falling of the tug-steamer, the lighter, and the boat – the turning of the windlass – the coming of the tide – that I myself seemed, to my own thinking, anything but new to the spot. Yet, I had never seen it in my life, a minute before, and had traversed two hundred miles to get at it. That very morning I had come bowling down, and struggling up, hill-country roads; looking back at snowy summits; meeting courteous peasants well to do, driving fat pigs and cattle to market: noting the neat and thrifty dwellings, with their unusual quantity of clean white linen, drying on the bushes; having windy weather suggested by every cotter's little rick, with its thatch straw-rigged and extra straw-ridged into over-lapping compartments like the back of a rhinoceros. Had I not given a lift of fourteen miles to the coast-guardsman (kit and all) who was coming to his spell of duty there, and had we not just now parted company? So it was; but the journey seemed to glide down into the placid sea, with other chafe and trouble, and for the moment nothing was so calmly and monotonously real under the sunlight as the gentle rising and falling of the water with its freight, the regular turning of the windlass aboard the lighter, and the slight obstruction so very near my feet.

O reader, haply turning this page by the fireside at home, and hearing the night wind rumbling in the chimney, that slight obstruction was the uppermost fragment of the wreck of the *Royal Charter*, Australian trader and passenger ship, homeward-bound, that struck here on that terrible morning of the twenty-sixth of this October, broke into three parts, went down with her treasure of at least five hundred human lives, and has never stirred since!

From which point, or from which, she drove ashore, stern foremost; on which side, or on which, she passed the little island in the bay, for ages henceforth to be aground certain yards outside her; these are rendered bootless questions by the darkness of that night and the darkness of death. Here she went down.

Even as I stood on the beach with the words 'Here she went down' in my ears, a diver in his grotesque dress, dipped heavily over the side of the boat alongside the lighter, and dropped to the bottom. On the shore by the water's edge, was a rough tent, made of fragments of wreck, where other divers and workmen sheltered themselves, and where they had kept Christmas day with rum and roast beef, to the destruction of their frail chimney. Cast up among the stones and boulders of the beach, were great spars of the lost vessel, and masses of iron twisted by the fury of the sea into the strangest forms. The timber was already bleached and iron rusted, and even these objects did no violence to the prevailing air the whole scene wore, of having been exactly the same for years and years.

Yet, only two short months had gone, since a man, living on the nearest hill-top overlooking the sea, being blown out of bed at about daybreak by the wind that had begun to strip his roof off, and getting upon a ladder with his nearest neighbour to construct some temporary device for keeping his house over his head, saw from the ladder's elevation as he looked down by chance towards the shore, some dark troubled object close in with the land. And he and the other, descending to the beach, and finding the sea mercilessly beating over a great broken ship, had clambered up the stoney ways, like staircases without stairs, on which the wild village hangs in little clusters, as fruit hangs on boughs, and had given the alarm. And so, over the hill-slopes, and past the waterfall, and down the gullies where the land drains off into the ocean, the scattered quarrymen and fishing men inhabiting that part of Wales had come running to the dismal sight – their clergyman among them. And as they stood in the leaden morning, stricken with pity, leaning hard against the wind, their breath and vision often failing as the sleet and spray rushed at them from the overforming and dissolving mountains of sea, and as the wool which was a part of the vessel's cargo blew in with the salt foam, and remained upon the land when the foam melted, they saw the ship's lifeboat put off from one of the heaps of wreck; and first, there were three men in her, and in a moment she capsized, and there were but two; and again, she was struck by a great mass of water, and there was but one; and again, she was thrown bottom upward, and that one, with his arm struck through the broken planks and waving as if for the help that could never reach him, went down into the deep.

It was the clergyman himself from whom I heard this, while I stood on the shore, looking in his kind wholesome face as it turned to the spot where the boat had been. The divers were down then, and busy. They were 'lifting' today the gold found yesterday – some five-and-twenty thousand pounds. Of three hundred and fifty thousand pounds worth of gold, three hundred pounds worth, in round numbers, was at that time recovered. The great bulk of the remainder was surely and steadily coming up. Some loss of sovereigns there would be, of course: indeed, at first sovereigns had drifted in with the sand, and been scattered far and wide over the beach, like sea-shells; but most other golden treasure would be found. As it was brought up, it went aboard the tug-steamer, where good account was taken of it. So tremendous had the force of the sea been when it broke the ship, that it had broken one great ingot of gold deep into a strong and heavy piece of her solid ironwork: in which, also, several loose sovereigns that the ingot had swept in before it, had been found, as firmly embedded as though the iron had been liquid when they were forced there. It had been remarked of such bodies come ashore, too, as had been seen by scientific men, that they had been stunned to death, and not suffocated. Observation, both of the internal change that had been wrought in them, and of their external expression, showed death to have been thus merciful and easy. The report was brought while I was holding such discourse on the beach, that no more bodies had come ashore since last night. It began to be very doubtful whether many more would be thrown up, until the north-east winds of the early spring set in. Moreover, a great number of the passengers, and particularly the second-class women passengers, were known to have been in the middle of the ship when she parted, and thus the collapsing wreck would have fallen upon them after yawning open, and would keep them down. A diver made known, even then, that he had come upon the body of a man, and had sought to release it from the great superincumbent weight; but that, finding he could not do so without mutilating the remains, he had left it where it was.

The Sirens

Lawrence Binyon

I heard a trumpet from beyond the moon,
Piercing the ice-blue gulfs of air,
Cry down the secret waters of the world,
Under the far sea-streams, to summon there
The foundered ships, the splendid ships, the lost ships.
In their ribbed ruin and age-long sleep they heard,
Where each had found her shadowy burial bed,
Clutched in blind reef, shoal-choked or shingle-bound;
Heard from betraying isles and capes of dread
In corners of all oceans, where the light
Gropes faltering over their spilt merchandise:
And shapes at last were stirred
On glimmerless abysses' oozy floors
Known to the dark fins only and drowned eyes; –
Sunk out of memory, they had glided forth
Bound from cold rivers to the tropic shores,
Or questing up the white gloom of the North,
Or shattered in the glory of old wars,
The laden ships, the gallant ships, the lost ships!

I saw them clouding up over the verge,
Ghosts that arose out of an unknown grave,
Strange to the buoyant seas that young they rode upon
And strange to the idle glitter of the wave.
Magically re-builded, rigged and manned,
They stole in their slow beauty toward the land.
Mariners, O Mariners!
I heard a voice cry; Home, come home!
Here is the rain-fresh earth; leaf-changing seasons; here
Spring the flowers; and here, older than memory, peace
Tastes on the air sweet as honey in the honey-comb.
Smells not the hearth-smoke better than spices of India?
Are not children's kisses dearer than ivory and pearls?
And sleep in the hill kinder than nameless water
And the cold, wandering foam?
Dear are the names of home, I heard a far voice answer,
Pleasant the tilled valley, the flocks and farms, and sweet
The hum in cities of men, and words of our own kin.
But we have tasted wild fruit, listened to strange music;
And all shores of the earth are but as doors of an inn;
We knocked at the doors, and slept; to arise at dawn and go.
We split blood for gold, trafficked in costly cargoes,
But knew in the end it was not these we sailed to win;
Only a wider sea; room for the winds to blow, and a world to wander in.

The Sinking of the *Vasa*

Dorothy Clewes

But what of a ship that sank to its burial ground at the hour of its birth? Such a ship was the *Vasa*, a Swedish seventeenth-century man-o'-war, built to the order of King Gustav Adolphus II. Every glory that could be portrayed in decoration went into its building. Rarely before had such a proud and splendid ship been seen. Its tragic and untimely end went into the annals of Swedish history. It became a legend – and for one man, a dream.

Anders Franzen was the man with the dream. It was a dream that had persisted from his earliest childhood – that one day he might find a lost galleon. As he grew up no other books had interested him but those about the ships that had gone down at the time of Sweden's greatness. As he grew to manhood, the dream became an obsession. It filled his mind to the exclusion of everything else. He spent long summer holidays divided between the archives in the museums and talking to fishermen and sailors: questioning them, listening intently to the stories they told. He sailed from island to island with trawl nets and sounding tackle.

Now and again his searchings rewarded him with wrecks of old, sunken ships, but none that came up to the ships he envisaged in his dreams, to that one ship. One thing began to dawn on him, however: and that was that these old wooden ships had *rotted* away and not been *eaten* away by the teredo navalis which destroys the woodwork of ships submerged in normal sea-water. The teredo navalis is the death-watch beetle of the sea and nothing is safe from its voracious jaws. Could this be, Franzen wondered, because of the more brackish waters of the Baltic? It was a new and exciting thought. If the beetle was not attracted to ships in these waters, what might he not find? Perhaps the fabulous and fabled *Vasa*. As a boy he had never tired of hearing his father talk about it.

King Gustav Adolphus had commissioned the building of the *Vasa* early in the seventeenth century. It was to be the flagship of the great Imperial Fleet, a man-o'-war, a galleon worthy of such a high office. Into the building of her went the finest work of those days: carving of great splendour and intricacy, embellished with gold leaf: figureheads of lions, of knights and mermaids, carvings of gods. Her superstructure rose proud and high and was emblazoned with decorations. She was weighted with cannon and fully masted: a proud tribute to the Swedish peoples. At last she was finished and the day of launching was proclaimed a holiday so that all might take delight in the great event. Crowds lined the harbour and the crew were privileged to take their families on board. It was a great and splendid occasion. And then – disaster struck. Barely had the wind filled her sails than she capsized and sank without trace.

This was the ship whose image filled the secret recesses of Anders Franzen's mind whenever he thought of lost galleons. It was too much to think that he would find it, but hope was always there.

He teased his brain to study the old Swedish language so that he could read the ancient and dusty MSS of that time, looking for some hint or intimation of its whereabouts. Three hundred years back men wrote in a close and difficult hand, using strange and unfamiliar words and phrases, but the more Anders Franzen read the more he got into the feel and rhythm of the writing. The difficulties did not deter him, rather they were a challenge that drove him on.

One day when he least expected it he came upon a mention of the *Vasa*. 'She sank behind Lustholmen,' the MS said, 'this side of Blockhusudden by Danviken.' It did not pinpoint the position precisely, but the reference was helpful and spurred him on to greater efforts.

The next exciting find was a map of the harbour, one that had been plotted out in the eighteenth century by another wouldbe hunter of old ships. A place called Stadsgardskajen was marked with a cross, and hinted that on this spot the *Vasa* went down.

The years were passing. It was now 1955. His friends thought him crazy, others thought that he must have some secret knowledge of material gain that he pursued his obsession with such determination. They could not believe that it was pure love of, and devotion to, a hobby that had possessed him always.

And then, one day in 1956, he came across the original letter sent to King Gustav Adolphus by the Council of the Realm. It was written two days after the disaster. There, set out in careful detail, was the chronicle of what had happened and the exact place where she had foundered. The words in the old letter caught and held Anders Franzen in a grip of intoxication. '. . . and this past Sunday, as stated, shortly after evensong . . . the wind south-west and the weather calm . . . and when she came out of the bay by Tegelviken there came more wind into her sails . . . and came to Beckholmsudden, where she entirely fell on her side.'

Impossible to imagine how Anders Franzen must have felt as he read those words: the culmination of years of dreaming and hoping, of study and search were perhaps to be rewarded at last.

Now Franzen's search took on a more serious note. He made soundings in the harbour off the island of Beckholm. It was by no means a case of finding at first try, nor did he expect it to be. He worked on with infinite patience, sending down his instrument bore, not too disappointed when no contact was made. And then, one unforgettable day, he felt the instrument bite and bore into some hard surface. With bated breath he hauled it in. In the ferrule of the bore was wedged a chunk of pitch black oak. He told himself that he mustn't jump to conclusions. It could be that he had struck any one of innumerable pieces of ancient sunken wreckage – but each time the bore went down, so each time did it come up with another telltale piece of oak. He was no distance at all off the dry dock entrance. If the *Vasa* had sunk immediately after launching this is where she could be expected to lie. It had to be the *Vasa*. He carefully made a note of the position and went ashore.

He made in all haste to the Royal Diving School. The Commanding Officer listened to Franzen's story with increasing excitement. The school was full of trainee divers, they had to perform their routine tests somewhere, and it might as well be over Franzen's 'wreck'.

The launch put out and chief diver Per Edvin Falting was kitted out in his watertight suit which combined weighted boots and a belt hung with weights. Then the helmet was fitted over his head and securely screwed down to the suit.

How different was his descent from the descent of the men who, in that year of disaster, had tried to salvage the wrecked ship. Now Edvin Falting had three hundred years of accumulated scientific knowledge to draw from as well as far superior equipment. He could breathe fresh air through the pipeline from the compressor, there was a lifeline at his service, a telephone, and a submarine electric lamp.

Franzen's feeling can hardly be imagined as he waited and watched the air bubbles rippling to the surface from Falting's exit pipe. Then the shock of hearing his voice reporting progress came over the loudspeaker: 'I am now lowering myself down beside a wreck. It is very dark, but I can feel a square opening in the hull. Here's another – and another –'

The square openings had to be the gun-ports of an ancient galleon. It must be the long-buried flagship. It must be the *Vasa*.

The running commentary went on as Falting explored. He could now see the remains of the sterncastle, rising high out of its mud-locked bed. It had been damaged by the efforts of earlier salvage workers, and by the anchors of ships coming into the harbour, but it was still, unmistakably, the stern-castle. There were other lost properties, unconnected with the *Vasa:* anchors, grappling irons, some of them no doubt answerable for the damage to the ship as they had trawled the harbour bed for whatever purpose, tearing into the wood-work, entangling with cannon and masts indiscriminately. Falting, encumbered as he was in his heavy suit, clambered clumsily over grapples, beams, statues – and all the while the beam from his lamp revealed an Aladdin's cave of wonders sunk in the muddy graveyard of the harbour mouth. Muzzles of ancient cannon leered at him from lower deck gun-ports, the wooden walls towering above. There was no longer any shadow of a doubt. This was the *Vasa* – the proud ship that so long ago, to the echoing acclamations of the crowd, had slipped her moorings to embark on a triumphant voyage of conquest. So long ago, and so different a picture now! Nevertheless whatever sad and sorry condition she was in, she was still the *Vasa*.

Over the loudspeaker the awed listeners heard the diver's voice taking formal possession of the ship on behalf of the crown.

In the full flush of such a thrilling discovery it would have been natural, understandable, if salvage had began at once, but cool heads prevailed. This was a project too vast and too important to attempt without great thought and care. Also, time and enormous expense had to be considered. A *Vasa* committee was formed. After deep deliberation and after every detail of salvage was gone into in the minutest particular they were at last ready to issue this order: 'RAISE THE *VASA!*'

One of the main values of the discovery of the *Vasa* was that until that time – the first half of the seventeenth century – hardly anything was known about naval architecture. As a serious craft it did not seem to exist. Added to which, the *Vasa*, agreed to be the oldest completely recognizable galleon in the world, going down as she had done at that time, had taken down with her to her watery grave, a priceless collection of articles of that period. In the words of Philippe Diole, the French marine archaeologist: 'An ancient shipwreck is a complete inventory and the most crystal-clear concentrated report concerning the culture at the time of the disaster. It is only on board ship, where so many testifying things are gathered together in a small space, that an entire cross-section of life can suddenly be arrested and preserved by the sea.'

The sea may have preserved the contents of the *Vasa*, but it was the immense weight of mud and clay which had preserved the woodwork of the ship herself – that and the absence of the teredo navalis, the voracious sea-beetle. Anders Franzen's first thoughts about the beetle had proved to be correct, they did not attack wooden ships in brackish waters. The question now was would she stand the enormous strain of being lifted? She had gone down in the full flowering of her magnificence, she had suffered no destruction by fire, no enemy had attacked her with gun-fire, no storm had ravaged her. For over three hundred years she had rested on the silting, salving mud and clay, and near enough to the naval dockyard to make the raising of her at least convenient as long as it was possible. The experts were only too well aware of the herculean task she presented.

What was more, while the mud and waters had preserved her, exposure to the air would quickly destroy her. Rapid drying would cause her to disappear before their eyes. Some way must be found to preserve her on land as well as she had been preserved under water. Some cover must be built for her where humidity could be controlled at all times – but that was in the future.

By great good fortune the Neptun Salvage Company owned two enormous pontoons. They had already served their original purpose – raising a Russian battleship sunk in the Baltic. Now they were to be used on a more heroic mission. Steel cables were to be passed under the *Vasa* and then attached to the two pontoons. In this way the *Vasa* would be cradled by the cables and then gently raised, but before that it would be necessary to dig six tunnels under her keel.

1957 saw the start of a new diving programme. The *Vasa* was first to be cleared of all movable objects and a careful inventory made of each piece, even if it was only a broken piece of ship's timber. In turn these had to be submerged in water to prevent drying out. When they ran out of water tanks, old household tubs were used.

On 9 September 1957, tunnelling began. Suction pipes were positioned and the mud gently drawn away. Now the mud began to give up its treasures. One of the first things to come to light was one of the gun-port lids. The mud had preserved it so well that some of the original gold leaf was still there. Then the magnificently carved mask of a lion's head broke surface. The finds were considered a good omen, and work went ahead with renewed enthusiasm.

One of the pieces of equipment which proved of inestimable service in the *Vasa's* recovery was the Zetterstrom-jet. This gave a pressurized stream of water which forced and washed away in one process the more solid mud. When the suction pipes had drawn away sufficient of the softer mud a diver descended into the prepared hole taking with him the Zetterstrom-jet. There, in only the dimmest light, a team of divers worked in relays of fifteen minutes each. In this way, very gradually, the mud was loosened. It was dangerous work, too, as at any moment who could tell whether the mud wall might not collapse and bury them alive? Everything possible was done to make the operation easier, the diver's lamps gave off a brilliant light, but so little could be seen from behind the glass window of the helmets. In great discomfort they battled on, one man's stint giving way to the next. And the deeper they bored the harder became the clay, so that the work got slower and slower. Winter caught up with them and work for that year had to be abandoned.

It was April before they could start the monotonous relays again, but at least they came to the ship's great keel – and now the process had to start all over again from the other side. In this way they bored the first tunnel. There were five more to go!

In between work on the tunnels the first bronze cannon was recovered from its gun port. If any further proof was needed that this ship was indeed the *Vasa*, the letters on the cannon proved its identity beyond all doubts. G.A.R.S. – Gustav Adolphus Rex Sueciae. Its weight was 2,500 pounds, scarred by anchors and trawling chains, but intact. Other finds were made: a part of the great main-mast; the tail of the figurehead lion; a carving of Hercules, symbol of strength; Cerberus, the hell dog; the oak bracket which supported the bowsprit. On another occasion a coat of arms, with cherubs supporting the *Vasa* sheaf – red paint and gold leaf still visible on the carving.

The finding of a keg of butter, and the opening of it after more than three hundred years was another thrill. It was tasted – and pronounced only slightly rancid! Beer barrels were more disappointing when it came to examining the liquid sloshing about inside – not even the telltale smell of beer remained. But the brandy cask proved its worthiness and was actually drinkable!

By the end of July 1959 the last tunnel was finished. Guide wires were passed under the keel of the *Vasa*. She was ready to be raised.

Now the pontoons came into action, taking up their positions with the salving vessels and the diving rafts. The pontoons were anchored either side of the line of rafts, then the pumps were started and the pontoons filled with water. Now the guide wires were replaced by heavy six-inch steel hawsers. They were going to take the full weight of the first lift.

By the end of August the men were ready to pump the water out of the pontoons. The rose very gradually out of the water, and the hawsers took over. Mud and clay fought to hold on to the prize it had held fast for nearly three hundred and fifty years but the steel hawsers and the pull of the pontoons were equal to the battle. One of the divers went down to see what progress was being made and reported that she had risen two feet from her muddy grave. On went the pumping and higher rose the pontoons. Now work began on sealing her gun-ports to make them water-tight, and the holes where iron nails had rusted away were plugged. There was no relaxation, the underwater repairing went on, plugging and caulking, until it was felt safe to start pumping the water out. She had been moved into more shallow water so that frogmen could now take over from the helmet divers: with their ease of movement the work speeded up. More leaks were found and the caulking went on. At first, as fast as water was pumped out of the *Vasa*, so did more water rush in to take its place, but gradually the great pumps began to win the battle against the inrush of water. She had now to be manoeuvred into the narrow dock where the *Vasa* committee planned to build a little house around her. Only in this way could humidity be controlled so that there would be no drying-out of her timbers, a process that would have to be continued over a period of years. So well had the job of pumping been carried out that the *Vasa* was able to float to the dock on her own keel. Sirens blew and flags fluttered in the breeze over the heads of the crowds lining the harbour walls – and was there, perhaps, an echo of the cheers of that other day so many years ago? Scarred and maimed, like a warrior returning from an ancient battlefield, the *Vasa* slid into her final resting place.

And so ended Anders Franzen's dream, more magnificently concluded than even he could have believed possible.

The Old Ships

James Elroy Flecker

I have seen old ships sail like swans asleep
Beyond the village which men still call Tyre,
With leaden age o'ercargoed, dipping deep
For Famagusta and the hidden sun
That rings black Cyprus with a lake of fire;
And all those ships were certainly so old –
Who knows how oft with squat and noisy gun
Questing brown slaves or Syrian oranges,
The pirate Genoese
Hell-raked them till they rolled
Blood, water, fruit, and corpses up the hold.
But now through friendly seas they softly run,
Painted the mid-sea blue or shore-sea green,
Still patterned with the vine and grapes in gold.

But I have seen
Pointing her shapely shadows from the dawn
And image tumbled on a rose-swept bay
A drowsy ship of some yet older day;
And, wonder's breath indrawn,
Thought I – who knows – but in that same
(Fished up beyond Aeaea, patched up new
– Stern painted brighter blue –)

That talkative, bald-headed seaman came
(Twelve patient comrades sweating at the oar)
From Troy's doom-crimson shore,
And with great lies about his wooden horse
Set the crew laughing, and forgot his course.
It was so old a ship – who knows, who knows?
– And yet so beautiful, I watched in vain
To see the mast burst open with a rose,
And the whole deck put on its leaves again.

The Kingdom of the Helmet

William Beebe

William Beebe found the joy of life 'half a mile down' into this new world. He was entranced by all he saw there, intoxicated by the strange beauty of it. He felt at home there – 'a returning native rather than an intruding stranger'. He longed to explore the deeper regions of the sea denied him by the limitations of the helmet with its attendant air-line to the surface. Later, working with Mr Otis Barton, he was to enjoy the greater freedom of the bathysphere they invented – the forerunner of Jacques Piccard's bathyscaph. In its confined space – fifty-four inches in diameter – he had his first glimpse of the wonders of the depths. Here, however, he describes those first early impressions seen from 'the kingdom of the helmet'.

When first I put on a diving helmet and climbed down the submerged ladder, then I knew that I had added thousands upon thousands of wonderful miles to my possible joy of earthly life: let me escape from dry-land etymology and say instead – the joys of planetary life; for personal exploration under the ocean is really unearthly – we are penetrating into a new world.

After we have dived hundreds of times we learn to discount the fears upon which we have been nurtured since childhood. And when the needless terrors of being water-enclosed, of the imputed malignity of octopuses, sharks, and barracudas have ceased to trouble our supreme delight in the strangeness and unbelievable beauties of this newly conquered realm, then we begin to appreciate the real significance of our achievement.

To enter into and to enjoy this new phase of life requires no practice or rehearsal, no special skill or elaborate preparation. If one dives and returns to the surface inarticulate with amazement and with a deep realization of the marvel of what he has seen and where he has been, then he deserves to go again and again. If he is unmoved or disappointed, then there remains for him on earth only a longer or shorter period of waiting for death; there can be little worthwhile left for him.

Ten years of diving on New York Zoological Society expeditions have taught me all the primal necessities. The only requirements are a bathing-suit and a pair of rubber-soled shoes, a copper helmet with glass set in front, an ordinary rubber hose, and a small hand pump. A folding metal ladder is excellent, but a rope is quite sufficient. Now you go into two, four, six, eight fathoms, swallowing as you descend to offset the increase of pressure. If your ears hurt severely a few feet below the surface ascend at once and go to the nearest aurist, for something is wrong and should be attended to, whether you ever dive again or not.

Forty feet is a good limit to set, and indeed the most brilliant and exciting forms of shore and reef life will be found in shallower depths. There is no danger of falling. If you stumble over the edge of a submerged cliff or lofty terrace, you simply half drift, half float, gently to the bottom. But when you stand on the edge of a deep chasm, and are already eight or ten fathoms down, don't let any alluring shell or coral lure you much deeper. Ears cannot withstand too great pressure.

After you have made a dozen descents you will wish to do something more than stand amazed, or vainly try to catch fish which swim close to the glass and look in at you.

As we have done, you can begin to devise all sorts of new apparatus. You wish to make notes, so get sheet zinc or pads of waterproof paper, find a comfortable block of coral and write as easily as if you were sitting in the boat. Be sure to tie your pencil tightly around, for otherwise the wood will separate and float to the surface, while the core of lead sinks to the bottom, to be nibbled at excitedly by small fry.

Motion pictures can be taken, down to twenty or twenty-five feet, by placing the camera in a tight brass box with a bit of glass in front. If you wish to paint, weight your easel with lead, waterproof your canvas or skin, and sit down with your palette of oils. You will have to brush away small fish from time to time, for some of the paints give forth an alluring odour and your palette will sometimes be covered with a hungry school of inchlings.

If you take your seat in the midst of a coral reef you may be attacked – not by giant octopuses or barracudas, or sharks – don't give them a thought – but you may feel a faint nip or a push at your elbow and there is a little fish – a demoiselle shorter than your thumb, all azure and gold, furiously butting at you. Her home is nearby and in its defence she fears nothing which swims, crawls or dives. Soon she will accept you as a harmless new kind of sea creature, and off she goes to drive away an approaching snapper or surgeonfish.

If your tastes incline to sport, invent submarine sling-shots and crossbows and shoot what particular fish you wish with barbed arrows of brass wire. I now use dynamite caps on the end of a weighted fishpole, but sling-shots and stabbing grains are safer for the beginner.

If you wish to make a garden, choose some beautiful slope or reef grotto and with a hatchet chop and pry off coral boulders with waving purple sea-plumes and golden sea-fans and great parti-coloured anemones. Wedge these into crevices, and in a few days you will have a sunken garden in a new and miraculous sense. As birds collect about the luxuriant growths of a garden in the upper air, so hosts of fish will follow your labours, great crabs and starfish will creep thither, and now and then fairy jellyfish will throb past, superior in beauty to anything in the upper world, more delicate and graceful than any butterfly.

Our grandmothers lined their garden paths with conch shells, but undersea it is more difficult to do this, for the giant snails will insist on walking away as soon as you have planted them. But other exquisite shells can be scattered about, and the easiest and quickest way to discover these is to search until you have found the hiding-place of an octopus, and here you will be certain to find a collection of empty shells of all kinds. The octopus is an adept at searching out toothsome molluscs, and he then carries them to his lair and devours the inmates at his leisure. The shells, quite perfect, are then thrown outside into his kitchen-midden.

Finally, as a border to your marine plantation, collect a score of small, rounded brain corals all thickly covered with tube worms. When you lay them in place, they will be a drab, dirty white. It is their momentary winter, but wait patiently and in five minutes you can see spring approach, and a host of pastel buds appear; and in another five minutes full summer arrives and your ivory mounds are ablaze with scarlet, mauve, blue, yellow, and green animal blossoms. All are in motion, though there is no current, and we feel there would be nothing remarkable in their suddenly saying, like Alice's tiger-lily: 'We *can* talk, when there's anybody worth talking to.'

The wise diver will refrain from written descriptions of his experiences. What I have published of under-sea-scapes has aroused commendation on the part of the fireside and dry land readers. The moment, however, one of them puts on a helmet and goes to see for himself, thereafter all words and

phrases, similes, and superlatives will become for him hopelessly inadequate. Just as the colours under-sea are nameless in the gamut of terrestrial hues, so our language becomes thin and vague when we try to fashion from it adequate submarine imagery. Even the commonest fishes and other organisms of our shallows are like different creatures when viewed from their element and their own level, instead of from man's vertical height above water: our human friends as we see them from a second-storey window are strangely unlike them face to face!

The Kingdom of the Helmet is not only a new experience for us, but the place in past aeons of time of fiercest competition and most spectacular evolution. It is a ribbon of a kingdom, of negligible depth – from six to sixty feet, and narrow in width – from a few inches to two miles. Its length is amazing – perhaps one hundred and fifty thousand miles of winding, submarine paths, rimming the rocks and cliffs of temperate fiords and bays; all along the palm-lined shores of southern continents, and the innumerable circles and rings of tropic isles and atolls. Perhaps the most interesting and exciting places are the reefs and shallows far from shore, like those of Bermuda which I have named Almost Island, where one can go overboard and to the bottom surrounded on all sides by depths forbidden to present ambulatory exploration.

When the summer's sun has warmed our northern waters, let us climb down the ladder off some rocky coast, say off Maine or Massachusetts. At once we begin to realize our new-found superiority: yesterday we crept painfully over legions of barnacles and peered ineffectually into outer depths. Now we pass quickly beyond the barnacle zone below lowest tide-mark, where things have been wet since creation. A little further down and the last of the steel-blue mussels passes from view, and then we perceive the great clinging roots of the giant seaweeds, with their leathery fronds stretching up and up to the surface. Green urchins give place to other larger species, and two or three fathoms down we enter the home of the beautiful basket starfish, hinting of the crinoids which have now almost vanished from earth.

We take our seat upon a mat of seaweed and watch the life of mid-water. Shrimps come in great numbers, drifting past like ghosts of living beings; the first squid, seen head on, will never be forgotten, nor will a galaxy of ctenophora jelly fish when the sunlight sets their cilia ablaze. Whelks and small curious crabs clamber upon our canvas shoes, and suddenly a thousand comets dash past – a school of herrings in search of spawning grounds. Only an impatient jerk on the hose will remind us that we have long overstayed our allotted time.

At the first dive in the tropics, say in the West Indies, we are impressed by the great increase in the amount of life and the unbelievable brilliancy of colour. Off New York we perhaps picked up a tiny crumb-of-bread sponge and on a clam shell found a bubble of coral the size of a marble. Here, in the midst of a tropical reef, corals form boulders six and eight feet across, or branched arborescent growths into which we climb. Anemones and fish are rainbow-tinted – harlequin angelfish and large-eyed scarlet squirrels. Horny corals send up unearthly purple branches like nothing conceivable above water, and the joy of it all is that everything that moves has little or no fear of us. We are made to feel at home – returning natives, not intruding strangers.

When many dives have been made at one place, so that the seascape has become familiar, and individual fish are known on sight and can be claimed as friends, then is the time to come out late some starlit evening, and go down in the dark. Choose a night when there is strong phosphorescence, and climb down the ladder very slowly. When your eyes pass just below the level of the water the illumination of the ripples is beyond any mere man-made fire. At first, as we stand on the bottom we seem to be in utter darkness, with only a glow coming down from above. A glance upward shows the keel of the boat turned to molten silver, and now our eyes have become re-adapted and our individual cosmos begins to be filled with galaxies and constellations, meteors and comets of blue and white light. These in turn are resolved by our intelligence into definite organisms. Some of them, such as jelly-fish and sea-worms, have lights of their own, but most of these shallow-water forms are illuminated by proxy. Every move they make evokes brilliance from the minute Noctiluca and other microscopic creatures. Now and then the passage of some great fish lights up all the surrounding reef with its caves and waving fronds, and memory, from our diurnal dives, supplies a host of details. Again, language fails us utterly; we can only stand and look and feel and later remember enough of the marvel of it all to wish to experience it again as soon as possible.

This is the Sea

Edward Shanks

This is the Sea. In these uneven walls
A wave lies prisoned. Far and far away,
Outward to ocean, as the slow tide falls,
Her sisters, through the capes that hold the bay,
Dancing in lovely liberty recede.
But lovely in captivity she lies,
Filled with soft colours, where the waving weed
Moves gently, and discloses to our eyes
Blurred shining veins of rock, and lucent shells
Under the light-shod water, and here repose
Small quiet fish, and dimly glowing bells
Of sleeping sea-anemones that close

Their tender fronds and will not now awake
Till on these rocks the waves returning break.

Two Miles Down in the Mediterranean

Jacques Piccard and Robert S. Dietz

Some years after William Beebe's invention of the bathysphere, Jacques Piccard improved on it by inventing the bathyscaph. This was a free-moving capsule – not unlike the balloon Auguste Piccard, his father, had invented in order to study cosmic rays in the stratosphere. Jacques Piccard intended his to operate in another element – the sea. It was to be a submarine for taking man to the unexplored depths, where no man had been before. In any invention there must always be the first trial, and this story tells of that.

The first of August 1953, a beautifully fine day, has special significance for me. That day the *Trieste* was launched, at Castellammare, culminating seven long months of final assembly. The Navalmeccanica's huge crane lifted her out and deposited her gently in the water. As she hung motionless in mid-air, many curious onlookers doubtlessly suppose that this zebra-striped blimp was intended to fly rather than dive. Her tank filled with air, the *Trieste* floated high in the water. Ballasting with nine tons of iron shot and 25,000 gallons of gasolene ensued. She now settled deeper and deeper until there remained almost no freeboard. Lying nearly awash, everyone could now see she was not a blimp, but a bathyscaph.

Eleven days later, my father and I made the baptismal test submersion. It was hardly a dive, as the bottom was only five fathoms down. Some engineers watching from the dock expressed surprise to see her sink and rise with such delicate precision and balance; the weight of a few pounds on the superstructure was enough to make the difference. But what could be simpler? The laws of physics are immutable; errors can be made in calculation and in design but the *Trieste's* were precise. She was a good ship.

Two days later we made another test harbour descent, doubling our depth this time. The bottom was littered with rusted chains, old anchors, and miscellaneous scrap iron. A pod of newsmen met us upon emersion and assaulted us with questions. 'What have you seen? Any phosphorescent fish? Treasure chests? Amphoras? Drowned Roman cities?' This last question was not too far-fetched, for Castellammare occupies the ancient site of Stabiae, overwhelmed along with Herculaneum and Pompeii by the A.D. 79 eruption of Vesuvius. It was near here that the elder Pliny met his death by approaching the eruption too closely.

'No,' we replied, 'just an old dead shell.'

This was too dull for good copy, so the next day we discovered in the papers that we had seen myriads of diverse shells and phosphorescent fishes.

Next it was in order to stage an open-sea test with a tug and full topside crew as a full-dress rehearsal. This final training dive, before towing out to Capri and Ponza, was to twenty-two fathoms. We landed in a beautiful marine garden. The current swept us slowly over the bottom for a thousand feet or more. We drifted past a large sea anemone and some other interesting forms of sea life.

Storms and rough seas set in, delaying us until the twenty-fifth of that month when we departed for Capri and the first dive. My father and I would continue to make the dives and take all the risks until the *Trieste's* inherent safety was amply demonstrated. Our confidence in the solidity of the cabin, the plastic windows, and the watertightness of wire lead-throughs into the cabin were such that we considered an unmanned dive unnecessary. It was too early to attempt serious oceanography; tests and trials were needed. A balloonist makes six ascents under the instructor's guidance before he is finally licensed. A bathyscaph pilot is on his own, there was no prior art.

The tow to Capri proved slow and tedious, for the *Trieste* had a tendency to yaw – to diverge from a straight line if towed too fast, zigzagging first to port and then to starboard. Provisionally, we fixed a sea anchor to the stern, resulting in a noticeable improvement.

The *Fenice*, a fast Italian navy corvette, joined us from Naples to watch over the diving area. When we rise to the surface, we must be sure that no small boat will ram us and pierce our thin-skinned float. There was no way of knowing exactly where the *Trieste* might break surface. Like a tireless sheep-dog, the *Fenice* quartered the sea, clearing a circle three miles across. One large steamer insisted on staying on course until a threat of force was issued. A speedboat carrying a celebrated movie-star also refused to yield its close view of the dive. Soft words proving vain, the *Fenice's* hoses went into action thoroughly dousing the craft. It moved off.

At grey first light we made ready to dive. This first attempt to dive deep was destined to meet with discouragement and to be involuntarily aborted. To begin with, a heavy shock overpowered the magnetically closed catch on the guide rope so that our 770-pound guide rope sank to the bottom. This guide rope is like a balloon's trail rope; when we approach the bottom it touches first and gradually lightens the craft, easing our touching down. There was a spare guide chain on board but its function was not vital. We decided to dive without it.

Then a scuba diver swam under the float to remove the screw caps from the ballast orifices. (The next year we installed a system which allows opening and closing the valves without the use of a diver.) My father and I climbed down into the cabin. Diving orders were relayed topside over the telephone.

'Pull out the pins.'

'Pins out.'

'Are you certain the pins are pulled?' Not to remove them could prove fatal: these pins are the safety ones for the ballast tubs so that they can be dropped in an emergency.

'I repeat, the pins are out. I am holding both of them here in my hands.'

'Good. Close the hatch and flood the air tanks. We are ready to dive.'

'. . . You are going down . . . the deck is awash. The tower is half under . . . we are disconnecting the telephone. *Buona fortuna!*'

The descent began. No sooner had we left the surface than we could see that the forward ballast tub was emptying itself. Our efforts to energize the solenoid and stop the black stream of iron pellets from pouring out were to no avail. A suba diver when removing the caps had apparently snagged his tank and broken the critical wire to the solenoid. This wire must have a weak point so as to part easily if the ballast tubs are ever jettisoned. After a few moments beneath the Mediterranean and sinking only a few score feet, we ignominiously rose back to the surface – such is the fail-safe security of the *Trieste*.

The possibility of returning to Castellammare, pumping out the gasoline, dry-docking and making repairs faced us. The diving season was already far advanced; the weather was deteriorating. We could not afford useless delays. Useless? Yes, there was a simple, satisfactory solution. We mechanically blocked the faulty ballast chute and refilled the tub with ballast. We would dive using only the aft ballast tub, which would throw us somewhat out of trim, but not seriously. Safety would not be diminished; we could always drop the faulty tub in its entirety if an emergency necessitated drastic action.

The rest of the morning was lost to re-ballasting after

capping the faulty tub. In the afternoon, we prepared to dive once again; this time events proceeded flawlessly. We settled slowly into the abyss, with the cabin lights extinguished. The blue light flooding in through the two four-inch portholes was sufficient to dimly illuminate the cabin. Gradually the shadows grew darker as the portholes shone like two large grey eyes, then dark grey, and finally black. Peering through the ports, we could see luminescent flashes; and, with the searchlight on, small white particles streaming by – the 'sea snow' – which would become a familiar sight to me and to every bathyscaph diver. A single, slightly phosphorescent fish about six inches long flashed momentarily into view, only to be gone in a trice. We were diving fast.

The recording depth gauge logged our descent: 200, 250, 400 and 500 fathoms. For this dive, we did not yet have a tachometer to show continuously our rate of descent; we could only compute it from the depth gauge graph. We also lacked an echo sounder for watching the illusion of the bottom 'falling-up' to meet us. In an area where the depth notations on the charts were not accurate, we were forced to resort to watch for the bottom through the porthole.

'. . . Five hundred fathoms', the searchlight streamed down into emptiness. Suddenly a dim, circular surface appeared in the cone of light – the bottom. I called to my father to brace himself for a rough landing. '*Tiens toi.*'

With insufficient time to jettison ballast, we struck bottom. So soft was the ooze that we felt no shock. I glanced at our depth gauge: it read 595 fathoms, or 3,570 feet. We tried to look out of the porthole hoping to see a pristine and fascinating vista. What did we see? Nothing! Mud was spattered against the porthole. (By later examination of the sphere we could tell that it had plunged 4·5 feet into the ooze.)

After fifteen minutes on the bottom making some technical observations, we decided it was useless to prolong our stay. I turned the switch to the working ballast tub, and waited: in

theory, the ballast should have been pouring out but there was no way to see and it pours out slowly. Several minutes elapsed, then the *Trieste* suddenly lurched forward and we were rising. My father rushed to the porthole in the hope of gaining a fleeting glimpse of the sea floor. Alas, in pulling out of the mud, the cabin stirred up a cloud of mud. By the time it cleared away, the bottom was already out of sight.

The ascent proceeded smoothly. The *Trieste* did not oscillate although our rate [racing yacht] reached 3·3 feet a second, and we were somewhat out of trim. A child's balloon rises in zigzags as will also a tennis ball released at the bottom of a swimming pool. This instability occurs when the object is rising too fast and creating strong eddies in its wake; slow it down and it will rise smoothly. It was difficult to determine for certain by calculation if the *Trieste* would remain below the critical speed, but my father thought it would. His optimism was justified, for we could not detect the slightest oscillation.

As we approached the surface, darkness first gave way to grey light, then growing into a limpid blue. A gentle bump followed by a slight rocking indicated our return to the surface some forty-five minutes after leaving bottom.

Menfish

Jacques Yves Cousteau

In 1945, with the second World War reaching its climax and France still an occupied country, Jacques Yves Cousteau perfected the first aqualung and the underwater world became a wonderland free to all who cared to venture there. From the early days when men had dived, holding their breath for the short period they could remain underwater, and at only limited depths, diving had undergone slow changes: the diving-bell, the cumbersome copper headpiece. Jacques Cousteau dreamed of a freedom that would allow men to explore the fascinating underwater world with the ease of fish. The age of the aqualung presaged a great new era of discovery, recovery and adventure. Man could now emulate the fish in his own element. This story tells of the first thrilling experience.

One morning in June 1943, I went to the railway station at Bandol on the French Riviera to collect a wooden case expressed from Paris. In it was a new and promising device, a result of years of struggle and dreams: an automatic compressed-air diving lung conceived by Emile Gagnan and myself. I rushed it to Villa Barry where my diving comrades, Philippe Tailliez and Frederic Dumas, were waiting. No children opened a Christmas present with more excitement than we did when we unpacked the first 'aqualung'. If it worked, diving would be revolutionized.

We found an assembly of moderate-sized cylinders of compressed air, linked to an air regulator the size of an alarm clock. The bottles contained air condensed to one hundred and fifty times atmospheric pressure. From the regulator there extended two tubes, joining on to a mouthpiece. With this equipment harnessed to the back, a watertight glass mask over the eyes and nose, and rubber foot fins, we intended to make unencumbered flights in the depths of the sea.

We hurried to a sheltered cove which would conceal our activity from curious bathers and Italian occupation troops. I checked the air pressure. It was difficult to contain my excitement and discuss calmly the plan of the first dive. Dumas, the best goggle diver in France, would stay on shore keeping warm and rested, ready, if necessary, to dive to my aid. My wife, Simone, would swim out on the surface with a schnorkel breathing-tube and watch me through her submerged mask. If she signalled anything had gone wrong, Dumas could dive to me in seconds. 'Didi', as he was known on the Riviera, could 'skin dive' to sixty feet.

My friends harnessed the three-cylinder block on my back with the regulator riding at the nape of my neck and the hoses looped over my head. I spat on the inside of my shatter-proof glass mask and rinsed it in the surf, so that mist would not form inside. I moulded the soft rubber flanges of the mask tightly over forehead and cheekbones and then fitted the mouthpiece under my lips and gripped the nodules between my teeth. A vent the size of a paperclip was to pass my inhalations and exhalations beneath the sea. Staggering under the fifty-pound apparatus I walked with a Charlie Chaplin waddle into the sea.

The diving lung was designed to be slightly buoyant. I reclined in the chilly water to estimate my compliance with Archimedes' principle that a solid body immersed in liquid is buoyed up by a force equal to the weight of the liquid displaced. Dumas justified me with Archimedes by attaching seven pounds of lead to my belt. I sank gently to the sand. I breathed sweet, effortless air. There was a faint whistle when I inhaled and a light rippling sound of bubbles when I breathed out. The regulator was adjusting pressure precisely to my needs.

I looked into the sea with the same sense of trespass that I have felt on every dive. A modest canyon opened below, full of dark green weeds, black sea urchins, and small flower-like white algae. Fingerlings browsed in the scene. The sand sloped down into a clear blue infinity. The sun struck so brightly I had to squint. My arms hanging at my sides, I kicked the fins languidly and travelled down, gaining speed, watching the beach reeling past. I stopped kicking and the momentum carried me on a fabulous glide. When I stopped, I slowly emptied my lungs and held my breath. The diminished volume of my body decreased the lifting force of the water, and I sank dreamily down. I inhaled a great chestful and retained it. I rose towards the surface.

My human lungs had a new rôle to play, that of a sensitive ballasting system. I took normal breaths in a slow rhythm, bowed my head and swam smoothly down to thirty feet. I felt no increasing water pressure, which at that depth is twice that of the surface. The aqualung automatically fed me increased compressed air to meet the new pressure layer. Through the fragile human lung linings this counter-pressure was being transmitted to the bloodstream and instantly spread throughout the incompressible body. My brain received no subjective news of the pressure and I was at ease except for a pain in the middle ear and sinus cavities. I swallowed, as one does when landing in an aeroplane, to open my eustachian tubes and stopped the pain. (I did not wear ear plugs, a dangerous practice when under water. Ear plugs would have trapped a pocket of air between them and the eardrums. Pressure building up in the eustachian tubes would have forced my eardrums outward, eventually to the bursting point.)

I reached the bottom in a state of excitement. A school of silvery sars (goat bream), round and flat as saucers, swam in a rocky chaos. I looked up and saw the surface shining like a defective mirror. In the centre of the looking-glass was the trim shilhouette of Simone, reduced to a doll. I waved. The doll waved at me.

I became fascinated with my exhalations. The bubbles swelled on the way up through lighter pressure layers, but were peculiarly flattened like mushroom caps by their eager push against the medium. I realized the importance bubbles were to have for us in the dives to come: as long as air bubbled on the surface all was well below. If the bubbles disappeared, there would be anxiety and emergency measures. They roared out of the regulator and kept me company. I felt less alone.

I swam across the rocks and compared myself favourably with the sars. To swim fishlike, horizontally, was the logical method in a medium eight hundred times denser than air. To halt and hang attached to nothing, no lines or air pipe to the surface, was a dream. At night I had often had visions of flying by extending my arms as wings. Now I flew without wings. (Since that first aqualung flight, I have never had a dream of flying.)

I thought of the helmet diver arriving on his ponderous boots at such a depth as this and struggling to walk a few yards, obsessed with his umbilici and his head imprisoned in copper. On 'skin-dives' I had seen him leaning dangerously forward to make a step, clamped in heavier pressure at the ankles than the head, a cripple in an alien land. From this day forward we would swim across miles of country no man had known, free and level, with our flesh feeling what the fish scales know.

I experimented with all possible manoeuvres of the aqualung – loops, somersaults, and barrel rolls. I stood upsidedown on one finger and burst out laughing – a shrill distorted laugh. Whatever I did, nothing altered the automatic rhythm of air. Delivered from gravity and buoyancy I flew around in space.

I could attain a speed of almost two knots without using my arms: I soared vertically and passed my own bubbles; I went down to sixty feet. We had been there many times without breathing aids but we did not know what happened below that boundary. How far could we go with this strange device?

Fifteen minutes had passed since I left the little cove. The regulator lisped in a steady cadence in the ten-fathom layer and I could spend an hour there on my air supply. I determined to stay as long as I could stand the chill. Here were tantalizing crevices we had been obliged to pass fleetingly before. I swam inch by inch into a dark, narrow tunnel, scraping my chest on the floor and banging the air tanks on the ceiling. In such situations a man is of two minds. One urges him on towards mystery and the other reminds him that, if he will use it, he is a creature with good sense that can keep him alive. I bounced against the ceiling. I had used one-third of my air and was getting lighter. My brain complained that this foolishness might sever my air hoses. I turned over and hung on my back.

The roof of the cave was thronged with lobsters. They stood there like great flies on a ceiling. Their heads and antennae were pointed towards the cave entrance. I breathed lesser lungsful to keep my chest from touching them. Above the water was occupied, ill-fed France. I thought of the hundreds of calories a diver loses in cold water. I selected a pair of one-pound lobsters and carefully plucked them from the roof. Without touching their stinging spines, I carried them towards the surface.

Simone had been floating, watching my bubbles wherever I went. She swam down towards me. I handed her the lobsters and went down again as she returned to the surface. She came up under a rock which bore a torpid Provençal citizen with a fishing rod. He saw a blonde girl emerge from the combers with lobsters wriggling in her hands. She said: 'Could you please watch these for me?' and put them on the rock. The fisherman dropped his rod.

Simone made five more surface dives to take lobsters from me and carry them to the rock. I surfaced in the cove, out of the fisherman's sight. Simone claimed her lobster swarm. She said: 'Keep one for yourself, monsieur. They are very easy to catch if you do as I did.'

Lunching on the treasure of the dive, Tailliez and Dumas questioned me on every detail. We revelled in plans for the aqualung. Tailliez pencilled the tablecloth and announced that each yard of depth we claimed in the sea would open to mankind three hundred thousand cubic kilometres of living space. Tailliez, Dumas and I had come a long way together. We had been eight years in the sea as goggle divers. Our new key to the hidden world promised wonders.

The Slimy Bottom of the Deep

William Shakespeare

From *King Richard III*, *Act I*, *Scene IV*

Methought that I had broken from the Tower,
And was embark'd to cross to Burgundy;
And in my company my brother Gloucester,
Who from my cabin tempted me to walk
Upon the hatches: thence we looked toward England,
And cited up a thousand heavy times,
During the wars of York and Lancaster,
That had befall'n us. As we pac'd along
Upon the giddy footing of the hatches,
Methought that Gloucester stumbled; and in falling,
Struck me, that thought to stay him, overboard
Into the tumbling billows of the main.
Lord, Lord! methought what pain it was to drown:
What dreadful noise of water in mine ears!
Methought I saw a thousand fearful wracks;
A thousand men that fishes gnawed upon;
Wedges of gold, great anchors, heaps of pearl,
Inestimable stones, unvalu'd jewels,
All scattered in the bottom of the sea.
Some lay in dead men's skulls; and in those holes
Where eyes did once inhabit, there were crept,
As 'twere in scorn of eyes, reflecting gems,
That wooed the slimy bottom of the deep,
And mocked the dead bones that lay scatter'd by.

Cave Diving

Jacques Yves Cousteau

Not all dives have the same idyllic quality and all diving does not take place in calm and predictable waters. Perhaps an element of danger enhances its appeal. There was a mystery surrounding the inland water cave of the Fountain of Vaucluse. Why did it flood so amazingly for just those few weeks in the year and after that drain away to a mere trickle? Hydrologists over hundreds of years had pondered over the strange phenomenon. Jacques Cousteau, at the suggestion of Commandant Brunet, went down with his friend and co-diver Frederic Dumas, to explore and to try to come to some conclusion.

Our worst experience in five thousand dives befell us not in the sea but in an inland water-cave, the famous Fountain of Vaucluse near Avignon. The renowned spring is a quiet pool, a crater under a six-hundred-foot limestone cliff above the river Sorgue. A trickle flows from it the year round, until March comes when the Fountain of Vaucluse erupts in a rage of water which swells the Sorgue to a flood. It pumps furiously for five weeks then subsides. The phenomenon has occurred every year in recorded history.

The fountain has evoked the fancy of poets since the Middle Ages. Petrarch wrote sonnets to Laura by the Fountain of Vaucluse in the fourteenth century. Frederic Mistral, a Provençal poet, was another admirer of the spring. Generations of hydrologists have leaned over the fountain, evolving dozens of theories. They have measured the rainfall on the plateau above, mapped the potholes in it, analysed the water, and determined that it is invariably 55° Fahrenheit all the year round. But no one knew what happened to discharge the amazing flood.

One principle of intermittent natural fountains is that of an underground syphon, which taps a pool of water lying higher inside the hill than the water level of the surface pool. Simple overflows of the inner pool by heavy rain seeping through the porous limestone did not explain Vaucluse, because it did not entirely respond to rainfall. There was either a huge inner reservoir or a series of inner caverns and a system of syphons. Scientific theories had no more validity than Mistral's explanation: 'One day the fairy of the fountain changed herself into a beautiful maiden and took an old strolling minstrel by the hand and led him down through Vaucluse's waters to an underground prairie, where seven huge diamonds plugged seven holes. "See these diamonds?" said the fairy. "When I lift the seventh, the fountain rises to the roots of the fig tree that drinks only once a year." ' Mistral's theory, as a matter of fact, possessed one more piece of tangible evidence than the scientific guesses. There is a rachitic hundred-year-old fig tree hooked on the vertical wall at the waterline of the annual flood. Its roots are watered but once a year.

A retired Army officer, Commandant Brunet, who had settled in the nearby village of Apt, became an addict of the Fountain as had Petrarch six hundred years before. The Commandant suggested that the Undersea Research Group dive into the Fountain and learn the secret of the mechanism. In 1946 the Navy gave us permission to try. We journeyed to Vaucluse on August 24, when the spring was quiescent. There seemed to be no point in entering a violent flood if its source might be discovered when the Fountain was quiet.

The arrival of uniformed naval officers and sailors in trucks loaded with diving equipment started a commotion in

Vaucluse. We were overwhelmed by boys, vying for the privilege of carrying our air cylinders, portable decompression chamber, aqualungs, and diving dresses, up the wooded trail to the Fountain. Half the town, led by Mayor Garcin, stopped work and accompanied us. They told us about the formidable dive into the Fountain by Señor Negri in 1936. He seemed to have been a remarkably bold type, for we were informed that he had descended in a diving suit with a microphone inside the helmet through which he broadcast a running account of his incredible rigours as he plunged one hundred and twenty feet to the lower elbow of the siphon. Our friends of Vaucluse recalled with a thrill the dramatic moment when the voice from the depths announced that Señor Negri had found Ottonelli's zinc boat!

We already knew about Negri and Ottonelli, the two men who had preceded us into the Fountain, Ottonelli in 1878. We greatly admired Ottonelli's dive in the primitive equipment of his era. We were somewhat mystified by Señor Negri, a Marseille salvage contractor, who had avoided seeing us on several occasions when we sought first-hand information on the topography of the Fountain. We had read his diving report, but we felt deprived of the details he might have given us personally.

The helmet divers described certain features to be found in the Fountain. Ottonelli's report stated that he had alighted on the bottom of a basin forty-five feet down and reached a depth of ninety feet in a sloping tunnel under a huge triangular stone. During the dive his zinc boat had capsized in the pool and slid down through the shaft. Negri said he had gone to one hundred and twenty feet, to the elbow of a syphon leading uphill, and found the zinc boat. The corrosion-proof metal had, of course, survived sixty years immersion. Negri reported he could proceed no further because his air pipe was dragging against a great boulder, precariously balanced on a pivot. The slightest move might have toppled the rock and pinned him down to a gruesome death.

We had predicated our tactical planning on the physical features described by the pioneer divers. Dumas and I were to form the first *cordée* – we used the mountain climber's term because we were to be tied together by a thirty-foot cord attached to our belts. Negri's measurements determined the length of our guide rope – four hundred feet – and the weights we carried on our belts, which were unusually heavy to allow us to penetrate the tunnel he had described and to plant ourselves against currents inside the syphon.

What we could not know until we had gone inside the Fountain was that Negri was over-imaginative. The topography of the cavern was completely unlike his description. Señor Negri's dramatic broadcast was probably delivered just out of sight of the watchers, about fifty feet down. Dumas and I all but gave our lives to learn that Ottonelli's boat never existed. That misinformation was not the only burden we carried into the Fountain: the new air compressor with which we filled the breathing cylinders had prepared a fantastic fate for us.

We adjusted our eyes to the gloom of the crater. Monsieur Garcin had lent us a Canadian canoe, which was floated over the throat of the Fountain, to anchor the guide rope. There was a heavy pig-iron weight on the end of the rope, which we wanted lowered beforehand as far as it would go down. The underwater entry was partially blocked by a huge stone buttress, but we managed to lower the pig-iron fifty-five feet. Chief Petty Officer Jean Pinard volunteered to dive without a protective suit to attempt to roll the pig-iron down as far as it was possible. Pinard returned lobster-red with cold and reported he had shoved the weight down to ninety feet. He did not suspect that he had been down further than Negri.

I donned my constant-volume diving dress over long woollens under the eyes of an appreciative audience perched round the rocky lip of the crater. My wife was among them, not liking this venture at all. Dumas wore an Italian Navy frogman outfit. We were loaded like donkeys. Each wore a three-cylinder lung, rubber foot fins, a heavy dagger, and two large waterproof flashlights, one in hand and one on the belt. Over my left arm was coiled three hundred feet of line in three pieces. Dumas carried an emergency micro-aqualung on his belt, a depth gauge, and a *piolet*, the alpinist's ice axe. There were rock slopes to be negotiated: with our heavy ballast we might need the *piolet*.

The surface commander was the late Lieutenant Maurice Fargues, our resourceful equipment officer. He was to keep his hand on the guide line as we transported the pig-iron down with us. The guide rope was our only communication with the surface. We had memorized a signal code. One tug from below requested Fargues to tighten the rope to clear snags. Three tugs meant pay out more line. Six tugs was the emergency signal for Fargues to haul us up as quickly as possible.

When the *cordée* reached Negri's syphon, we planned to station the pig-iron, and attach to it one of the lengths of rope I carried over my arm. As we climbed on into the syphon, I would unreel this line behind me. We believed that our goal would be found past Negri's see-sawing rock, up a long sloping arm of the syphon, in an air cave, where in some manner unknown Vaucluse's annual eruption was launched.

Embarrassed by the wealth of gadgets we had hanging on to us, and needing our comrades' support, we waded into the pool. We looked around for the last time. I saw the reassuring silhouette of Fargues and the crowd round the amphitheatre. In their forefront was a young abbé, who had no doubt come to be of service in a certain eventuality.

As we submerged, the water liberated us from weight. We stayed motionless in the pool for a minute to test our ballast and communications system. Under my flexible helmet I had a special mouthpiece which allowed me to articulate under water. Dumas had no speaking facility, but could answer me with nods and gestures.

I turned face down and plunged through the dark door. I rapidly passed the buttress into the shaft, unworried about Dumas's keeping pace on the thirty-foot cord at my waist. He can outswim me any time. Our dive was a trial run: we were the first *cordée* of a series. We intended to waste no time on details of topography but to proceed directly to the pig-iron and take it on to the elbow of Negri's syphon, from which we would quickly take up a new thread into the secret of the Fountain. In retrospect, I also find that my subconscious mechanism was anxious to conclude the first dive as soon as possible.

I glanced back and saw Didi gliding easily through the door against a faint green haze. The sky was no longer our business. We belonged now to a world where no light had ever struck. I could not see my flashlight beam beneath me in the frightening dark – the water had no suspended motes to reflect light. A disc of light blinked on and off in the darkness when my flashlight beam hit rock. I went head down with tigerish speed, sinking by my overballast, unmindful of Dumas. Suddenly I was held by the belt and stones rattled past me. Heavier borne than I, Dumas was trying to brake his fall with his feet. His suit was filling with water. Big limestone blocks came loose and rumbled down round me. A stone bounced off my shoulder. I remotely realized I should try to think. I could not think.

Ninety feet down I found the pig-iron standing on a ledge. It did not appear in the torch beam as an object from the world above, but as something germane to this place. Dimly I recalled that I must do something about the pig-iron. I shoved it down the slope. It roared down with Dumas's stones. During this blurred effort I did not notice that I had lost the lines coiled on my arm. I did not know that I had failed to give Fargues three tugs on the line to pay out the weight. I had forgotten Fargues and everything behind us. The tunnel broke into a sharper decline. I circled my right hand continuously, playing the torch in spirals on the clean and polished walls. I was travelling at two knots. I was in the Paris subway. I met nobody. There was nobody in the Metro, not a single rock bass. No fish at all.

At that time of year our ears are well trained to pressure after a summer's diving. *Why did my ears ache so?* Something was happening. The light no longer ran around the tunnel walls. The beam spread on a flat bottom, covered with pebbles. It was earth, not rock, the detritus of the chasm. I could find no walls. I was on the floor of a vast drowned cave. I found the pig-iron, but no zinc boat, no syphon, and no precariously balanced rock. My head ached. I was drained of initiative.

I returned to our purpose, to learn the geography of the immensity that had no visible roof or walls, but rolled away down at a forty-five-degree incline. I could not surface without searching the ceiling for the hole that led up to the inner cavern of our theory.

I was attached to something, I remembered. The flashlight picked out a rope which curled off to a strange form floating supine above the pebbles. Dumas hung there in his cumbersome equipment, holding his torch like a ridiculous glow-worm. Only his arms were moving. He was sleepily trying to tie his *piolet* to the pig-iron line. His black frogman suit was filling with water. He struggled weakly to inflate it with compressed air. I swam to him and looked at his depth gauge. It read one hundred and fifty feet. The dial was flooded. We were deeper than that. We were at least two hundred feet down, four hundred feet away from the surface at the bottom of a crooked slanting tunnel.

We had rapture of the depths, but not the familiar drunkenness. We felt heavy and anxious, instead of exuberant. Dumas was stricken worse than I. I thought: *This is not how I should feel at this depth . . . I can't go back until I learn where we are. Why don't I feel a current? The pig-iron line is our only way home. What if we lose it? Where is the rope I had on my arm?* I was able in that instant to recall that I had lost the line somewhere above. I took Dumas's hand and closed it round the guide line. 'Stay here,' I shouted. 'I'll find the shaft.' Dumas understood me to mean I had no air and needed the safety aqualung. I sent the beam of the flashlight round in search of the roof of the cave. I found no ceiling.

Dumas was passing under heavy narcosis. He thought I was the one in danger. He fumbled to release the emergency lung. As he tugged hopelessly at his belt, he scudded across the drowned shingle and abandoned the guide line to the surface. The rope dissolved in the dark. I was swimming above, mulishly seeking for a wall or a ceiling, when I felt his weight tugging me back like a drifting anchor, restraining my search.

Above us somewhere were seventy fathoms of tunnel and crumbling rock. My weakened brain found the power to conjure up our fate. When our air ran out we would grope along the ceiling and suffocate in dulled agony. I shook off this thought and swam down to the ebbing glow of Dumas's flashlight.

He had almost lost consciousness. When I touched him, he grabbed my wrist with awful strength and hauled me towards him for a final experience of life, an embrace that would take me with him. I twisted out of his hold and backed away. I examined Dumas with the torch. I saw his protruding eyes rolling inside the mask.

The cave was quiet between my gasping breaths. I marshalled all my remaining brain power to consider the situation. Fortunately there was no current to carry Dumas away from the pig-iron. If there had been the least current, we would have been lost. *The pig-iron must be near.* I looked for that rusted metal block, more precious than gold. And suddenly there it was, stolid and reassuring. Its line flew away into the dark, towards the hope of life.

In his stupor, Didi lost control of his jaws and his mouthpiece slipped from his teeth. He swallowed water and took some in his lungs before he somehow got the grip back into his mouth. Now, with the guide line beckoning, I realized that I could not swim to the surface, carrying the inert Dumas; who weighed at least twenty-five pounds in his waterlogged suit. I was in a state of exhaustion from the mysterious effect of the cave. We had not exercised strenuously, yet Dumas was helpless and I was becoming idiotic.

I would climb the rope, dragging Dumas with me. I grasped the pig-iron rope and started up, hand-over-hand, with Dumas drifting below, along the smooth vertical rock.

My first three hand-holds on the line were interpreted correctly by Fargues as the signal to pay out more rope. He did so, with a will. With utter dismay I saw the rope slackening and made superhuman efforts to climb it. Fargues smartly fed me rope when he felt my traction. It took an eternal minute for me to work out the right tactics, namely that I should continue to haul down the rope, until the end of it came into Fargues' hand. He would never let go. I hauled the rope in dull glee.

Four hundred feet of rope passed through my hands and curled into the cave. And a knot came into my hands. Fargues was giving us more rope to penetrate the ultimate gallery of Vaucluse. He had efficiently tied on another length to encourage us to pass deeper.

I dropped the rope like an enemy. I would have to climb the tunnel slope like an alpinist. Foot by foot I climbed the finger-holds of rock, stopping when I lost my respiratory rhythm by exertion and was near to fainting. I drove myself on, and felt that I was making progress. I reached for a good hand-hold, standing on the tips of my fins. The crag eluded my fingers and I was dragged down by the weight of Dumas.

The shock turned my mind to the rope again and I suddenly recalled our signals: six tugs meant pull everything up. I grabbed the line and jerked it, confident that I could count to six. The line was slack and snagged on obstacles in the four hundred feet to Maurice Fargues. *Fargues, do you not understand my situation?* I was at the end of my strength. Dumas was hanging on me.

Why doesn't Dumas understand how bad he is for me? Dumas, you will die, anyway. Maybe you are already gone. Didi, I hate to do it, but you are dead and you will not let me live. Go away, Didi. I reached for my belt dagger and prepared to cut the cord to Dumas.

Even in my incompetence there was something that held the knife in its holster. *Before I cut you off, Didi, I will try again to reach Fargues.* I took the line and repeated the distress signal, again and again. *Didi, I am doing all a man can do. I am dying too.*

On shore, Fargues stood in perplexed concentration. The first *cordée* had not been down for the full period of the plan, but the strange pattern of our signals disturbed him. His hard but sensitive hand on the rope had felt no clear signals since the episode a few minutes back when suddenly we wanted lots of rope. He had given it to us, eagerly adding another length. *They must have found something tremendous down there,* thought Fargues. He was eager to penetrate the mystery himself on a later dive. Yet he was uneasy about the lifelessness of the rope in the last few minutes. He frowned and fingered the rope like a pulse, and waited.

Up from the lag of rope, four hundred feet across the friction of rocks, and through the surface, a faint vibration tickled Fargue's finger. He reacted by standing and grumbling, half to himself, half to the cave watchers, '*Qu'est-ce que je risque? De me faire engueuler?*' (What do I risk? Being sworn at?) With a set face he hauled the pig-iron in.

I felt the rope tighten. I jerked my hand off the dagger and hung on. Dumas's air cylinders rang on the rocks as we were borne swiftly up. A hundred feet above I saw a faint triangle of green light, where hope lay. In less than a minute Fargues pulled us out into the pool and leaped in the water after the senseless Dumas. Tailliez and Pinard waded in after me. I gathered what strength I had left to control my emotions, not to break down. I managed to walk out of the pool. Dumas lay on his stomach and vomited. Our friends stripped off our rubber suits. I warmed myself round a cauldron of flaming petrol. Fargues and the doctor worked over Dumas. In five minutes he was on his feet, standing by the fire. I handed him a bottle of brandy. He took a drink and said, 'I'm going down again.' I wondered where Simone was.

The Mayor said, 'When your air bubbles stopped coming to the surface, you wife ran down the hill. She said she could not stand it.' Poor Simone had raced to a café in Vaucluse and ordered the most powerful spirit in the house. A rumour-monger raced through the village, yelling that one of the divers was drowned. Simone cried, 'Which one? What colour was his mask?'

'Red,' said the harbinger.

Simone gasped with relief – my mask was blue. Then she thought of Didi in his red mask and her joy collapsed. She returned distractedly up the trail to the Fountain. There stood Didi, a miracle to her.

Dumas's recuperative powers soon brought his colour back and his mind cleared. He wanted to know why we had been drugged in the cavern. In the afternoon another *cordée*, Taillicz and Guy Morandicrc, prepared to dive, without the junk we had carried. They wore only long underwear and light ballast, which made them slightly buoyant. They planned to go the cavern and reconnoitre for the passage which led to the secret of Vaucluse. As soon as they found it, they would immediately return and sketch the layout for the third *cordée*, which would make the final plunge.

From the diving logs of Captain Tailliez and Morandiere, I am able to recount their experience, which was almost as appalling as ours. Certainly it took greater courage than ours to enter the Fountain from which we had been so luckily saved. In the few minutes they spent just under the surface of the pool, getting used to the water, Morandiere felt intense cold. They entered the tunnel abreast, roped together. Second *cordée* tactics were to swim down side by side along the ceiling.

When they encountered humps sticking down from the roof, they were to duck under them and then return to follow the ceiling closely. Each hump they met promised to level off beyond, but never did. They went down and down. Our only depth gauge had been ruined, but the veteran Tailliez had a sharp physiological sense of depth. At an estimated one hundred and twenty feet he halted the march so they might study their subjective sensations. Tailliez felt the first inviting throbs of rapture of the depths. He knew that to be impossible at a mere twenty fathoms. However, the symptoms were pronounced.

He called to Morandiere that they should turn back. Morandiere manoeuvred himself and the rope to facilitate Tailliez's turnabout. As he did so, he heard that Tailliez's respiratory rhythm was disorderly, and faced his partner so that Tailliez could see him give six pulls on the pig-iron rope. Unable to exchange words under water, the team had to depend on errant flashlight beams and understanding to accomplish the turn. Morandiere stationed himself below Tailliez to conduct the Captain to the surface. Tailliez construed these activities to mean that Morandiere was in trouble. Both men were slipping into the blank rapture that had almost fiinished the first *cordée*.

Tailliez carefully climbed the guide line. The rope behind drifted aimlessly in the water, and a loop hung round his shoulders. Tailliez felt he had to sever the rope before it entangled him. He whipped out his dagger and cut it away. Morandiere, swimming freely below him, was afraid his mate was passing out. The confused second *cordée* ascended to the green hall light of the Fountain. Morandiere closed in, took Tailliez's feet, and gave him a strong boost through the narrow door. The effort upset Morandiere's breathing cycle.

We saw Tailliez emerge in his white underwear, Morandiere following through the underwater door. Tailliez broke the surface, found a footing, and walked out of the water, erect and wild-eyed. In his right hand he held his dagger, upside down. His fingers were cut to the bone by the blade and blood was flowing down his sodden woollens. He did not feel it.

We resolved to call it a day with a shallow plunge to map the entrance of the Fountain. We made sure that Didi, in his anger against the cave, could not slip down to the drowned cavern that had nearly been our tomb. Fargues lashed a 150-foot line to Dumas's waist and took Didi's dagger to prevent him cutting himself loose and going down further. The final reconnaissance of the entrance shaft passed without incident.

It was an emotional day. That evening in Vaucluse the first and second *cordées* made a subjective comparison of cognac narcosis and rapture of the Fountain. None of us could relax, thinking of the enigmatic stupor that had overtaken us. We knew the berserk intoxication of *l'ivresse des grandes profondeurs* at two hundred and fifty feet in the sea, but why did this clear, lifeless limestone water cheat a man's mind in a different way?

Simone, Didi and I drove back to Toulon that night, thinking hard, despite fatigue and headache. Long silences were spaced by occasional suggestions. Didi said, 'Narcotic effects aren't the only cause of diving accidents. There are social and subjective fears, the air you breathe . . .' I jumped at the idea. 'The air you breathe!' I said. 'Let's run a lab test on the air left in the lungs.'

The next morning we sampled the cylinders. The analysis showed 1/2000 of carbon monoxide. At a depth of one hundred and sixty feet the effect of carbon monoxide is sixfold. The amount we were breathing may kill a man in twenty minutes. We started our new Diesel-powered free-piston air compressor. We saw the compressor sucking in its own exhaust fumes. We had all been breathing lethal doses of carbon monoxide.

Fish Talk

Hans Hass

The lighter side to diving is recounted by Hans Hass. Hans Hass is one of the great names in the early history of underwater-diving. Beginning as a skin-diving hunter, he soon discarded that frivolous pastime for the more serious pursuit of marine biology and photography. His deep commitment to the underwater world has made him wonder about the mystery of fish intelligence and communication and he has come to the belief that it may not be so silent a world as we may think, and it is more likely to be we who are deaf to fish language. In a piece from his book *Welcome From the Sea*, he tells us how he arrived at this conclusion.

Many things in nature that we take for granted turn out when we get to closer grips with them to be very puzzling. There is, for instance, the question of how fishes find their way about. Since they have eyes, it is pretty obvious that they can observe their surroundings. That there is a problem, however, is at once evident if we put on a diving mask and go down under the turbid waters of a large river such as the Danube or the Mississippi.

You can barely see four inches in front of you, yet fishes live in such conditions. They swim about, they seek their food, they chase after other fish, are themselves hunted, they get together in shoals and remain together, they seek out their mates, they have a hole in which they live and to which, despite the current, they always find their way back. They do all this, although they can hardly see anything, and they never collide with a stone or rock. How do they manage it?

Then, again, what happens at night? It is well known that many carnivorous fish hunt for preference at night-time. Sharks swim at night through inky blackness, steer their course with incomparable skill through coral reefs, track down their prey, chase after it – and never bump into anything.

I got a hint of a possible solution to the puzzle during my very first attempt at diving off the coast of the French Rivieria. I took my harpoon with me and swam in clear water illuminated with brilliant sunshine. But the fish did not seem to see me with their eyes as to feel my movements. If I made a clumsy plunge and smacked about on the surface, then every fish in the neighbourhood had got a warning. If, when I was stalking one, I made too brusque a movement, then at once he would dart off. When, however, I swam silently and easily then I could get so close to many fish that I was within striking distance of them – although their eyes were turned right in my direction.

Once I was swimming close beside a steep wall of rock against which fairly large waves were breaking when I saw, a good way off and about six fathoms below the surface, the tail of a larger grouper sticking out from behind a stone. The creature was almost hidden and only the slowly and gently moving tail was visible. Still looking downwards I extricated myself from a cloud of foam and splashed to the surface. There was the fish whirling round and round. Its head appeared and the creature peered up at me. There was no doubt at all that he had noticed my presence. Despite the thundering impact of the waves, he had heard the slight noise I had made by my clumsy movements.

We may be sitting at a concert. A thousand notes and sounds fill the air around us. Then a door creaks and despite the waves of music we hear that creak. Obviously we hear it with our ears, but we record the noise with our brains. So it may have been with that grouper. The pounding of the surf was something he was accustomed to, but the noise of my movements was for him like the creaking of a door. It is true that fish have no external auditory meatus, but such an opening is not necessary in the water as the tissues themselves are admirable conductors of sound. Also, it would appear that the swim-bladder plays a part in picking up sounds. Fish can be trained to recognize certain notes, one for food, another for a blow with a glass stick, and they soon learn to distinguish between the sounds. Then, the notes will be struck at closer intervals. In this way it has been found out that carp, for instance, possess an acuity of hearing comparable with that of human beings, so that sounds differing from one another by a quarter of a note can be recognized.

Now we come to the problem as to whether musical sounds can be compared with those made by a swimmer in the water, that is to say whether fishes really do 'hear' with their ears such vibrations through the water.

Fishes possess, in a number of regions of their body, small buds which occur in especial profusion along the sides of the animal. These sensitive protuberances are in connection with a canal filled with mucus and which through numerous openings is in contact with the surrounding water. From each sensitive bud there extends a diminutive hair-like sense-organ into the canal. Sharks have, in addition, peculiar sense-cavities in their heads, the so-called ampullae of Lorenzini. All these organs have been known for a long time, but no one has been very sure as to what their function is. Generally speaking, they have been ignored or explained away quite vaguely as organs for the perception of water currents. Since, however, the nerves from the sense-buds lead to certain well-developed areas in the fish brain, it must be assumed that the organs have some important significance for the animal.

In a few books I have found the suggestion advanced that the 'lateral lines' provide fish with a tactile sense that operates at a distance, just as is the case with bats which move about in the dark without ever colliding with anything. This peculiarity of bats' flight was a lasting puzzle until two researchers discovered that bats, when flying, emit rapid squeaks which are inaudible to the human ear. These cries are echoed back from surrounding objects (just like the rays from a pocket torch when we flash them onto anything in the dark) and the echo is picked up by the extraordinarily sensitive ears of the bat. Thus they find their way about and perceive their surroundings as though they were provided with radar. They recognize obstacles although they cannot see them with their eyes.

Maybe it is more or less the same with fish. The movements they make in swimming send out vibrations into the water and these vibrations may be returned to them from rocks and other objects, vibrations which may be perceived by the 'lateral lines' of the fish. Thus the creatures may be able to 'see' what lies around them even in quite turbid or dark waters, and 'see' not by means of rays of light but by the aid of the impact of vibrations which, sooner or later, reach the individual, sensitive observation-buds on the body and enable the animal to conceive in its brain a spatial, three-dimensional image of its environment.

All this sounds very complicated, I know, but perhaps only because we cannot easily imagine something that lies outside the sphere of our own senses. We are like opaque sacks pierced with a few holes. Through such holes – our senses – we see and perceive the world around us. But there are plenty of animals which have other 'holes' than ours, other senses, so that they perceive and judge the world in a way quite different from ourselves.

For instance, whereas for us sight is the principal sense, a dog 'sees' mainly with his nose. His picture of the world is in all probability not even conceivable to us. Still more remarkable is the case of many insects which possess some sense functions which are quite incomprehensible for us. The same may be true of fish and sharks. Since, in dark and turbid water, their eyes are only of limited use to them, such creatures may construct their image of the world, first and foremost, from the impressions which are communicated to them through the vibrations in the water.

Thus would be explained why fish and sharks, when they perceive a man, always put themselves sideways. This would be not so much to turn their eyes to a lateral position as to place their bodies so that the sensitive areas may receive the impressions conveyed by vibration through the water. The wider apart these sensitive areas lie from each other, so much the stronger would be the plastic image received by the animal – just as we use a stereo-telescope to increase the focal length of our eyes and so obtain a better perspective of distant objects. We may then come to the interesting conclusion that the bigger a fish or a shark (and thus the greater the area sensitive to vibrations) the better is its range of perception and its conception of space.

I had got as far as this in my theory when, in 1939, we made some very interesting and, I think, remarkable observations off the coast of the West Indian island of Curaçao.

On the stormy northern coast where there are many sharks these always put in an appearance when we harpooned a large fish; indeed, there would not elapse more than from ten to twenty seconds before the creatures could be seen, although just before there was not a sign of them in all the wide expanse of the sea. Had the smell of blood attracted them? Hardly. Since visibility under water was as much as forty yards, the sharks must have been at least fifty and possibly a hundred yards away. It was impossible that a smell of blood could spread so rapidly. So what? We came to the conclusion it must have been the frightened struggling and flapping about of the harpooned fish that they heard, or, to use the term we have just been employing, that they 'saw'.

Again, if we missed a fish and the creature in fright swirled around vigorously, then, at once, up came a shark. In this case there could be no question of the marauder being attracted by the smell of blood. At the first opportunity I dived, at the same spot, right down to the sea-bottom and flapped about as hard as I could with my flippers. We waited. No shark came. A little later we shot a fish and several sharks appeared at once.

Either my theory was false and it was not the noise of the struggling that attracted the sharks (though maybe the fish emit a cry that is inaudible to us) or sharks can distinguish between the splashing and floundering about of a desperate grouper and the noise made by my flippers. For the sharks there were two different sorts of vibrations.

This second possibility gave me some food for thought. Since the various sorts of fish are very different in shape and possess also very distinctively formed fins, it might be that each sort of fish emits into the water vibrations which are characteristic of and peculiar to it. Fish even of the same kind and species have different movements according to their state at any given moment. A healthy fish does not behave as does a sick fish, a hungry fish has a different comportment from a fish at mating time. Might it not be possible that sharks – and fishes, too – could perceive such differences? Could it be that, in this way, they might be able to recognize each other even in turbid waters and in the dark?

I remembered some fish we had seen swimming quite close past the snout of a shark without anything happening. Maybe they knew from the shark's movements that he was full up and not inclined for hunting.

Then there are fish in love play. We had noticed how, in many species, the partners swam round and round one another while rapidly vibrating their fins. Was such comportment just the expression of their excitement or was it just the opposite? It might be that fish by trilling their fins get themselves into a state for copulation. Maybe it is with their fins that they make their declarations of love, exactly as birds do with mating calls.

In the depths of the oceans there are many fish provided with luminous organs which may serve them to distinguish between the other fish they may meet. Still, it is not very probable that a hunted fish will light the way for its pursuer! In the depths there are robber fish which have to face many difficulties in tracking down and recognizing their prey. Maybe these creatures are guided by the melody of fins.

If these suppositions of mine are valid, then the silent world of the sea must be full of noise even if it is a noise made up of countless 'voices' just as imperceptible to our ears as are the cries emitted by bats in their flight. Even the black, terrifying abyss of the ocean would be quite 'light' and 'welcoming' though it is not illuminated by the sun's rays but lighted by vibrations that zigzag and crisscross through the moving waters.

The Fish

Rupert Brooke

In cool and curving world he lies
And ripples with dark ecstasies.
The kind luxurious lapse and steal

Shapes all his universe to feel
And know and be; the clinging stream
Closes his memory, glooms his dream,
Who lips the roots o' the shore, and glides
Superb on returning tides,
Those silent waters weave for him
A fluctuant mutable world and dim,
Where wavering masses bulge and gape
Mysterious, and shape to shape
Dies momently through whorl and hollow,
And form and line and solid follow
Solid and line and form to dream
Fantastic down the eternal stream;
An obscure world, a shifting world,
Bulbous, or pulled to thin, or curled,
Or serpentine, or driving arrows,
Or serene sidings, or March narrows.
There slipping wave and shore are one,
And weed and mud. No ray of sun,
But glow to glow fades down the deep
(As dream to unknown dream in sleep);
Shaken translucency illumes
The hyaline of drifting glooms;
The strange soft-handed depth subdues
Drowned colour there, but black to hues
As death to living, decomposes –
Red darkness of the heart of roses,
Blue brilliant from dead starless skies,
And gold that lies behind the eyes,
The unknown, unnameable sightless white
That is the essential flame of night,
Lustreless purple, hooded green;
The myriad hues that lie between
Darkness and darkness! . . .

And all's one

Gentle, embracing, quiet, dun,
The world he rests in, world he knows.
Perpetual curving. Only – grows
An eddy in that ordered falling,
A knowledge from the gloom, a calling
Weed in the wave, gleam in the mud —
The dark fire leaps along his blood;
Dateless and deathless, blind and still,
The intricate impulse works its will;
His woven world drops back; and he,
Sans providence, sans memory,
Unconscious and directly driven,
Fades to some dank sufficient heaven.

O world of lips, O world of laughter,
Where hope is fleet and thought flies after,
Of lights in the clear night, of cries
That drift along the wave and rise
This to the glittering stars above,
You know the hands, the eyes of love!
The strife of limbs, the sightless clinging,
The infinite distance, and the singing
Blown by the wind, a flame of sound,
The gleam, the flowers, and vast around
The horizon, and the heights above –
You know the sigh, the song of love!

But there the night is close, and there
Darkncss is cold and strangc and barc;
And the secret deeps are whisperless;
And rhythm is all deliciousness;
And joy is in the throbbing tide,
Whose intricate fingers beat and glide
In felt bewildering harmonies
Of trembling touch; and music is

The exquisite knocking of the blood.
Space is no more, under the mud;
His bliss is older than the sun.
Silent and straight the waters run.
The lights, the cries, the willows dim,
And the dark tide are one with him.

Salt Pond Bay

Robert E. Schroeder

Between receiving his master's degree and a doctorate, Robert E. Schroeder accepted a position on the U.S. Virgin Islands as a member of their Fisheries Research Project. The island he was to work from was St John, which is one of the smaller islands. It is a magical island of soft green hills and rain forests rising out of the deep blue of the Caribbean, fringed with coral reefs and silver sand. Inland, traces still remain of the old estate houses that belonged to the sugar planters, and the jungle growth has not altogether choked the ruins of the plantations where the slaves worked the old mills. One can still stumble upon a sugar kettle and the crumbling remains of storehouses, and the dungeons still wear their tell-tale iron bolts that held the slaves captive. Now it is a tropical paradise, its clear warm waters a perfect introduction to marine research. To make perfection more perfect, as well as solving the housing problem for himself, Robert Schroeder bought a boat and made that his headquarters, anchored in Lameshur Bay, on the south side of the island. Here he tells of his first experience of night diving in nearby Salt Pond Bay.

The ocean was oily black, and the night moonless. The shore, where the ripples softly washed the rocks, was blacker still, a silhouette against the stars. Fumbling in the dark, I slipped into the harness of my aqualung and fastened the weight-belt.

Jack (Dr John Randall), in charge of the research station,

lowered the long cord of our improvized diving light to the bottom, forty feet down. I put on my mask and leaned over the side. All I could see was a pale circle of sand and coral, pathetically small and far away in the deep dark ocean.

I cast off the little dinghy that carried the battery. A sudden splash told me that Jack was on his way, and a moment later his ripple-distorted figure appeared under the skiff, barely visible in the dim light. Gathering my courage I slid over the gunwale and the warm, dark Caribbean closed over my head.

The light seemed miles away, and we swam toward it down a long wavery tunnel of darkness, accompanied by excited finger-length shadow-fishes that darted about in the eerie glow. Jack grasped the light, and to my profound relief swung the beam about to reassure himself that nothing was lurking in the dark. No monsters! But the ruby-bright eyes of night-walking lobsters blazed from the corals, and pale fish swam over the sands. An ivory-white mantis shrimp at my feet lifted huge refractile eyes to stare blindly from its burrow, and a sand-dwelling anemone stretched gracefully curving tentacles of crimson and gold.

I had been diving in tropical waters for pleasure and science for several years and had collected reef fishes in the Bahamas and the Florida Keys, but this was my first experience with night diving. Strange creatures I had never seen on the sunny daytime reefs indulged in mysterious pursuits. Familiar organisms had so changed their colours and their habits that they were almost unrecognizable. My initial anxiety was quickly forgotten in the wonder of a scene transmuted by darkness. I realized I had been unaware of half the problems in the sea; that night was as important as day, and almost entirely different.

We soon found what we were looking for near the bay's mouth on a stretch of sand and coral rubble. Baby conch do indeed burrow by day to come out at night to feed. Jack swam about excitedly grabbing every one he could find; but to me

K

the conch were incidental to the wonders of the ocean at night.

Stream-lined blood-red worms, attracted by the light, swam around us in circles. Blotchy black and white goatfish, which school along reefs by day, had assumed coats of mottled scarlet and were sleeping on the sand. Scorpionfish, blinded by our light, allowed me to scoop them up in my palm. Cardinalfish, which by day are orange-red and hide deep in the coral, were translucent ghosts of themselves, swimming high above the reef. I dragged Jack away from his baby conch, appropriated the light, and went sightseeing.

A brick-red octopus the size of a grapefruit sat bolt upright and motionless on the coral rubble. Light-blinded, it did not move until Jack urged it into the mouth of his collecting bag. The creature crawled into the bag, revealing that it had been sitting on the overturned shell of a large queen conch, its tentacles and mantle completely enclosing the aperture.

The conch had withdrawn far into its shell, but did not appear the worse for its experience. Was the octopus waiting for it to emerge to be paralysed by a poisonous bite, or had a poison already been secreted into the enclosed shell cavity? The octopus is a flexible creature, and easily could have squeezed far enough into the shell to bite the occupant. Perhaps the siege had only begun.

The poisonous bite of the octopus may explain the presence of conch in the stomach of fish which have no means of removing the big snail from its shell. Jack once found both a conch and an octopus in a snapper's stomach. Had a snapper attacked the octopus after it paralysed the conch, it would have gotten both the octopus and its victim. Or the conch might have been left on the bottom to be scavenged by crabs or carnivorous snails, such as Murex, that are incapable of attacking a healthy conch.

We made several night dives at Salt Pond Bay. One night two small manta rays, perhaps ten feet in wing span, circled briefly above us in the glow of the light. I was afraid they would hit the extension cord, capsizing the battery boat and plunging us into darkness, but after two circuits they slipped out of the light and disappeared. Another night I was examining expanded coral polyps when Jack tapped me on the shoulder and pointed. Floating above us were four huge tarpon. I watched in awe as they hung mesmerized by the blinding glare before slowly drifting towards us, their expressionless eyes casting pale reflections.

Tarpon are among the more primitive bony fish. To me they always had an ancient and alien look, as if belonging to a bygone age. Their appearance this night in ghostly silver against the darkness did nothing to reduce this impression. It was as though we were diving on a fossil reef sixty million years ago, when mammals were still new.

I looked at Jack. He pointed at the nearest tarpon. I also pointed, questioningly, with my spear. We had collected no tarpon. Jack nodded. I assumed he meant we were to break our night-time rule against spearing except in self-defence.

Drawing a careful bead on the nearest five-footer, I fired. The spear rebounded, a great scale impaled on its tip. The tarpon disappeared straight up, and its companion zipped out of sight. I waited a moment, but failed to see where the first one broke the surface.

Jack and I both went up after that. In the skiff, Jean (Mrs Robert Schroeder) was calling at the top of her lungs: 'What was it? Bob, are you all right? What on earth . . . ?' We swam to the skiff, where she was jumping up and down excitedly. 'A great silvery thing! It soared out of the water like a rocket! I didn't see where it came down!' Handing her our equipment, we hauled ourselves aboard. 'It was a tarpon, a big one. We didn't see where it came down either. It went through the surface and just disappeared.' Jack looked at the sky speculatively and said wryly: 'maybe it went into orbit!'

A Cache of Silver

Roland Morris

In more recent years the underwater accent has been on the search for ancient and historical wrecks. Sophisticated electronic equipment has now made this possible. What could be more romantic than the discovery of a fabled man-o'-war that has lain in its ocean grave for three, four hundred years? What more intriguing than the finding of bronze cannon, silver plate, pieces of eight and gold dubloons? Echoes of well-loved stories come whispering back to us: the hunt for Captain Flint's treasure led by old blind Pew. How many of us have not wished ourselves into the shoes of Jim Hawkins?

Roland Morris did not wish himself in Jim Hawkins' shoes but rather he wished to live a real life treasure experience of his own. He is an ex-salvage diver and lives in Penzance. The Isles of Scilly are a few hundred oars-lengths from his home, The Admiral Benbow. He knows the islands well and is respectfully aware of the tides and currents that run there. He was also aware that fathoms deep in those waters lay the wreck of the *Association*, the flagship of Admiral Sir Cloudesley Shovell. She had been built in 1696. In March 1707 Sir Cloudesley Shovell and his fleet were ordered to rendezvous off the French coast prior to a planned attack on Toulon. A distinguished company sailed with him. From the outset they were bedevilled by bad weather – mist and fog made the usual observations difficult and gales forced them off-course time and time again. At last, confused as to their true whereabouts, Sir Cloudesley called a conference with the Captains of the rest of the fleet. There was some dissension but it was finally agreed that they were off Ushant and close to their point of rendezvous. They sailed on, unaware that they were heading for the treacherous approaches to the Isles of Scilly. The warning beacon was flashing from St Agnes, but either its light was misunderstood, or the foul weather obscured it. A few minutes later the *Association* had struck the Gilstone Rock and went down with all hands. The wreck of the *Association*, lying deep in these waters, fascinated Roland Morris – the period it conjured up as much as the wealth that must lie buried with it. His salvaging work at an end he applied for the rights to work on the wreck – along with other ships known to be wrecked there. He gathered together his divers and equipment and set off for the island. One of the divers who made up the team was Geoffrey Upton, a young crayfish diver from the mainland, another was Douglas Rowe, who Roland Morris had found already stationed at Scilly. Both were experienced divers. From his carefully studied chart Roland Morris knew, more or less, where the wreck should be – but they had to wait upon the weather.

In the snatches of comparative calm his diver located two large anchors and much iron shot. It was certain that it came from an ancient man-o'-war, but there was no positive proof that it came from the *Association:* that was to come later. From the position of the salvaged anchors they now felt that the ship must have gone down on the outer ledges of the Gilstone. On subsequent dives they located bronze cannon and on one occasion one of the divers came up with a gold coin. It was in mint condition and was dated 1706: a Portuguese 4,000-reis gold coin. They felt that they were getting warm. Many dives and several bronze cannon later Roland Morris wrote:

The sea was an incorruptible blue; it had a guiltless charm which would have fooled anyone who had not seen it over the last few weeks; it was as limpid and as smooth as a clear conscience. The Gilstone itself squatted there guileless and childlike, with the wavelets lapping at its sides like lambs about their mother. As for the Outer Ledge, the water on the upper reaches of the Thames could not have been smoother, more penitent, more contrite, then there. The sea plays a deep game at the Western Rocks – fathoms deep.

The day was one of sun, for the sun predominated; there was no wind, no rough sea, there were no clouds in the sky – so the sun took over. At noon-tide it beat down from a clear sky; its garish reflection pranced and darted across the sea in a wide stream of brilliance, flowing from the far-off hazy horizon, right up to our low-lying boat.

For week after week we had suffered the battering, the strain, of working on rough water; this calm was unbelievable; it was so strange, everything looked so neat and tidy. Nature could not have arranged things better – it was fitting to have fair weather on this day – our day of all days.

Geoff and Doug were both surfacing together, their bubbles had told me that they were coming up; I could see the colour of their yellow air bottles flickering way down in the depths, their brilliantly coloured hoods made streaks of orange in the limpid green water. They popped up side by side and handed up their underwater torches; together they took their demand valves from their mouths, together they uncoupled their bottle harness and one at a time lifted their 'sets' inboard, for this day I was their only 'tender'. They both sat at the side of the boat; in silence, except for their heavy breathing, they took off their face glasses, their gloves and flippers. I said nothing – there was nothing to say. I waited for them to speak. I noticed that they were both looking at me.

'You tell him,' Geoff said, presently, still looking at me.

'No, no – go on, you tell him,' said Doug; he also was looking at me. I thought this was odd, they were talking to each other but were looking fixedly at me.

'Why? What's happened?' I asked. 'Is everything all right?'

'Fine,' said Geoff, laughing at my concern. 'We've found a hell of a lot of coins, that's all!'

'You haven't!' I said, or something equally fatuous.

'Yes,' said Doug. 'There's thousands and thousands of them!' Each time he said 'thousands' he spread his arms out wide.

'The whole floor is covered with them,' said Geoff. 'We couldn't believe our eyes! I kept jabbing my torch in a different place, pointing out more and more coins to Doug – and he was doing the same to me.'

'Well, look,' said Doug, in his calm, business-like way, as if he had a train to catch, 'we've got enough air for another dive. I'll go down first while you tell Roland all about it, we can't both work in there together. I'll take that spike hammer down with me, I'll leave it there for you when I come up.'

'Take something with you to put the coins in,' said Geoff.

After Doug had disappeared with a splash I asked:

'Can't get in where together? What's he want the spike hammer for?' (A spike hammer is a slater's hammer with a long curving spike, ideal for underwater work.)

Jabbing his finger on the storm-boat's bottom boards on which we were sprawled, each jab to emphasize some point he was explaining, Geoff told me of their amazing discovery.

It seems that the two divers were exploring the submarine tunnels and caverns in close company. These passages and caves are formed, in the main, by the space between rocks and under rocks; rocks lying in great derangement, one upon another; rocks, some of them, which reach the size of a bungalow. The visibility was almost nil, so each of our divers was armed with an underwater torch. The going, at best, was slow; every inch of the terrain had to be inspected, every likely-looking hump or lump had to be tapped with a small mason's hammer to see what it consisted of. Geoff came on a small opening in the rocks into which he shone his torch, but the interior was so devoid of light it was impossible to see far in. However, it could be seen that the opening widened out again just inside the entrance. Although the opening was too small for a fully kitted-up diver to pass, Geoff decided – 'more or less on the spur of the moment' – to have a look inside. He unharnessed his air bottles from his back; pushing these in front of him with one hand, holding his torch with the other, he wormed his way – mainly by using his elbows – through the opening. It was a cave-like space under the tumbled rocks and had but one entrance. As he knelt on the floor of the 'cave', as he sorted himself out from his struggles in getting through the entrance, he felt the coins under the palms of his hands. They were stacked slantingly against each other, like slates on a roof. When he shone his torch on them, when he looked closely, he could see that the bottom was paved with coins, stuck together in a 'carpet' which stretched the full area of the floor space.

The cave had very little head-room, but, nevertheless, Geoff signalled for Doug to come in; Doug had watched Geoff's manoeuvres in gaining an entrance, so on all fours he followed suit. With both divers occupying the confined space, like a couple of excited animals fussing about in a pen, the fine succulent silt rose up from under them in a black cloud. Their torches were worse than useless in the circumstances, in the Stygian darkness. Doug decided to make an exit, but when he came to move his air bottles he found they were entangled with those of Geoff's. Patting each other on the back as if to say, 'Don't move, stay where you are until this lot settles,' they waited for the water to clear.

'There was a steady current passing through,' said Geoff, 'we watched the sediment flowing past the beams of our torches like smoke.'

The coins the divers found were well and truly stuck down. Here and there, glued and immovable among the coagulated mass, were many round shot, pouting up like bubbles of mud in a swamp.

After one or two tentative, but ineffectual, attempts to release a likely-looking lump from the parent mass, the divers gave up. They searched out several loose gold and silver coins, and put these to one side; but before they made their exit from the darkness of that hidey-hole they found it incumbent upon themselves to lay down their torches, and to execute – if I can so describe it – a sort of war-whoop of victory over the treasure trove. Squatting down on their haunches they jigged up and down on the silver coins, slapping the rough, fabulous 'carpet' with their flippers and the flats of their hands, all the time sucking away at their rubber umbilical tubes attached to their air supply which they held precariously in their laps. Oh, that I had been a barnacle of the wall of that rocky igloo! After their short, exhausting, frog-like dance, they waited while the suspended grit drifted away, shook hands with each other and wormed their way out of the narrow entrance.

Any exhilarated reaction of mine to the news of the find was tempered a little by a feeling of admiration for the divers, and also a sense of anxiety for their safety. Manoeuvres of the kind that Geoff had described to me call for the use of a 'surface demand' type of air supply. This is a supply of air which is carried in cylinders in the diving boat, and air in this case being taken, under pressure, from the surface down to the diver by a lengthy air line. Some divers do not mind this method – the only one I have ever experienced – other divers, owing to the tethering restriction of movement caused by the rubber pipe, will have little to do with it in any circumstances.

By the time Doug had come up with a large clump of blackened, encrusted pieces of eight and several golden coins, weighing several pounds in all, my words of praise had begun to resemble the blurb not unlike that which describes some epic of the screen: 'Stupendous! Fantastic! Sensational!' – and it was all of that.

With Doug up from his last dive of the day, Geoff dropped over the side for his third and final dive. I sat back against the sun-warmed timbers of the boat. I watched the ripples radiating out over the smooth water from the spot where Geoff had disappeared. I could not help but pause and wonder; just the depth of the sea away from us lay treasure; just seventy feet of water, but what an obstruction, what a prevention that had been over the years. Only today – after two hundred years – had that watery barrier between treasure and discovery been breached. And now, here in the bottom of the boat, lay slim gold coins and chunky silver pieces of eight.

Doug sat on the floor of the boat with his legs stretched out in front of him; with a pencil as a lever, he was prising the coins apart. In addition to the pieces of eight there were English crowns, half-crowns, and shilling pieces; the coins of the outer layer of the lump of silver were, naturally, not so well preserved as those on the inside, most of the latter being in perfect condition.

'Look at this,' Doug said. He washed a coin in the tin of salt water beside him and handed it over to me. It was a Charles II crown. 'DEL. GRATIA. – CAROLUS II.' it read; on the date side was the inscription: 'REX.MAG.BR.ET.HIB.1666'.

There was the proud King's profile – he resembled a Roman emperor complete with laureate head. It was in fine condition; there were many such crowns in the grimy lump of silver.

When Geoff came up he brought an even larger mass of coins.

'My goodness,' he gasped, 'that lot's going to take some getting out! We'll need a water jet or something to help us.' He dropped the silver lump with a thump on the bottom boards; I placed it in a bucket of sea-water so that the deposit should not harden further.

These were the first two treasure-lifting dives of the many which were to take place over the next six days, dives which occupied the divers from one and a half to two hours a day.

During the salvaging there was opposition from several quarters, amateur as well as professional, but at this time nothing could be done about it. Now legislation is on the way to protect the salvor who has paid good money for his wreck and the right to work on it for a set period – on the way, but not law as yet, so there will be predators waiting to move in after the labour of others, for some time yet.

It was some months later that proof beyond any shadow of a doubt was found that the wreck Roland Morris had found really was the *Association*. It came with the finding of a solid silver plate.

On it was a crest. It was a beautiful piece of work and it was in fine condition. Mr Morris was able at last to identify it as the crest of Sir Cloudesley Shovell: the arms of the Narborough family, Sir Cloudesley having married the widow of Admiral Narborough.

Toll for the Brave

William Cowper

Toll for the brave!
The brave that are no more!
All sunk beneath the wave,
Fast by their native shore!

Eight hundred of the brave,
Whose courage well was tried,
Had made the vessel heel,
And laid her on her side.

A land breeze shook the shrouds,
And she was overset;
Down went the *Royal George*
With all her crew complete.

Toll for the brave!
Brave Kempenfelt has gone;
His last sea fight is fought,
His work of glory done.

It was not in the battle,
No tempest gave the shock;
She sprang no fatal leak,
She ran upon no rock.

His sword was in its sheath,
His fingers held the pen,
When Kempenfelt went down,
With twice four hundred men.

Weigh the vessel up,
Once dreaded by our foes!
And mingle with our cup
The tear that England owes.

Her timbers yet are sound,
And she may float again
Full charged with England's thunder,
And plough the distant main.

But Kempenfelt is gone,
His victories are o'er;
And he and his eight hundred
Shall plough the wave no more.

Resurrection from the Sea

Dorothy Clewes

When one thinks of salvage from the sea one's mind inevitably turns to ships, but since inventive man conquered the air with his flying machines he also added another risk for mankind. Flying has now become a perfected art, with no more danger to life and limb than the risk we take every day crossing a road. In fact, nowadays, flying is many times safer than travelling on the soul-less motorways that are threading their way over more and more of our once green and pleasant land. Experienced air crews fly passengers millions of miles with no more concern than if they were taking them by bus to the nearest shopping centre. The BOAC Comet was the first of the jet planes. De Havilland's pride in their new aircraft was understandable and justified. They had logged many thousands of miles taking their passengers swiftly and in great comfort and safety to the four corners of the world. What then went wrong on that fateful Sunday of 10 January 1954 when Comet 1 G-ALYP, en route from Singapore to London, left Rome Airport at 09.31 GNT., and shortly afterwards came down in the sea near the island of Elba? This is the story of the unravelling of that mystery, of the salvaging of the many scattered pieces of wreckage, and of the assembly of them into a mock-up of the machine in an endeavour to discover the cause of its disintegration.

After an uneventful flight from Singapore the Comet landed at Rome's Ciampino Airport to refuel and change pilots. The take-over pilot took off for an equally uneventful flight to London. Now and again he reported his position to the control tower at Ciampino: '... heading for Ostia Beacon ... climbing 2,500 ft ...' – and a little later: '... Abeam Civita-vecchia Beacon, 19,000 ft ... visual flight above clouds at 23,000 ft ... Over Orbetello Beacon now ...'

A few minutes before, the pilot of an Argonaut which had also taken off from Rome had been having an R/T conversation with the Comet pilot, and asked for cloud-top height. When the Comet pilot came back from reporting his position to Ciampino Control Tower, he must have remembered he had not given the Argonaut pilot the cloud-top height he had asked for. He called him back: '... did you get my ...' That was all the Argonaut pilot heard: transmission had come to an abrupt end.

At Elba, a sailor busy at his work heard a heavy roaring

noise like thunder. He turned in the direction of the noise and saw a globe of fire rotating as it came down into the sea.

There were other witnesses: a farmer said he heard the noise of an aircraft but did not pay any heed to it. Suddenly, however, his attention was caught by a roaring sound in the air in the direction from which the noise of the aircraft was coming, and he distinctly noticed two pieces of an aircraft, the smaller piece in flames, falling in almost parallel lines into the sea.

A driver on the island said he suddenly heard a number of explosions and a great roaring noise. He turned in the direction from which the noise was coming and saw a red flame falling into the sea, followed by a wake of smoke in the form of a spiral . . .

The officer commanding the Harbour Authority at Portoferraio, Elba, went into action immediately and air and sea search and rescue operations were rushed to the spot. Bodies of fifteen of the victims were recovered. The distintegrated Comet had plummeted to the bottom of the sea with its crew of six and the rest of its twenty-nine passengers.

At this point one could only hazard a guess at what had happened: fire on board? an act of sabotage? – but it was only guesswork. There was one way, though, of getting to grips with the mystery – and that was to recover what could be salvaged of the Comet from its watery grave and build a mock-up to try and fit together the jigsaw events that had led up to the accident.

The great salvage operation got underway. In the meantime the bodies of the fifteen victims were examined. No evidence was found that they had been subject to the explosion of a bomb, neither had drowning been the cause of death. The marks found on the bodies were consistent with violent displacement and sudden impact with parts of the aircraft.

This being so, what had caused the break-up of the Comet?

The Admiralty put out an order for salvage operations to

be launched. The work was to extend over a period of nearly three months with Her Majesty's ships and Italian trawlers taking part. The area was scanned with electronic equipment, notes were taken of reports from eye-witnesses and of local fishermen who had been trawling at the time of the crash. Wreckage was in this way found lying on mud 400 feet below the surface. So far so good. The ships were loaded with search and salvage equipment including a TV camera. The Admiralty already had one of these cameras in use when it was discovered that the manufacturers had a new and more sophisticated one – but only as yet on the drawing-board. A massive effort was made by the manufacturers to get a prototype model built. They managed it within a week and flew it out with two of their engineers.

There followed weeks of painstaking search. Bad weather hampered the proceedings and disappointment dampened the spirits of the eleven men on board the salvage ship – but at last, in February, a piece of Comet wreckage was sighted by the TV camera. An observation chamber fed by oxygen, and with a man aboard, was lowered. The ship had to be moored securely for this operation but it presented no problem to the Italians who were experts in deep-sea diving. The salvage ship cruised dead slow many miles up and down the contact area. Nothing must upset the camera which was sweeping a path twelve foot wide. A twelve foot width is not much in the wide expanse of ocean and the ship had to go over the position time and time again.

And then things began to happen. Pieces of unmistakable evidence began to show up on the screen, nothing big, but enough to assure them that they were on the right track. Now the job could truly begin.

A crack diver went over the side and down the taut line to a piece of wreckage. Before the thrill of the find could be savoured, up he came and down went his relief. Now the grab had to be guided to a position to lift up the finds.

The haul had raised their hopes but nothing more came to light and as a final effort before moving on a trawl was laid – a very simple affair that circled an area and brought the ends back to the salvage ship – and it caught on something. At first it resisted all efforts to be moved, and the men on board were fearful that the trawl would break. Within twenty feet of the surface a frogman went down to investigate. The catch turned out to be a chair from the Comet, tangled up with some electric cables attached to the after pressure dome. A sling was put round the whole and carefully and slowly it was lifted up on to the deck. It brought up something else that hadn't been identified on the TV screen: the ladies and gents toilets.

All that week the grab worked on, bringing up mouthfuls of wreckage. It was like striking a seam of gold and they worked it until the vein ran out and until heavy seas once more forced them to leave the area.

It was March before they could take up the work again. The sea was calmer but there was still a heavy swell so that sometimes the TV camera was pitched too high and sometimes it was pitched so low that it got its face buried in the mud. In the middle of the month, after waiting for the weather to settle a diver was once more lowered in the observation chamber – and this time he struck really lucky. He found the largest piece of wreckage yet.

The grab was lowered and up came a rear girder – all 65 feet of it with jagged projections. Down went the grab again and this time hauled up an even larger front spar, added to which the diver had sighted two jet engines. A few days later the engines were lifted, then the starboard undercarriage, and miscellaneous objects like bits of a wing, flaps, cowlings, engine components, wireless sets, flight instruments – even someone's de luxe camera – and on another drop, the light-meter belonging to it.

It was some time before a find as good as that came again –

but when it did it was as if the sea had been saving up this as a kind of reward for gruelling, painstaking, dedicated labour. The diver in the observation chamber reported a really large piece of – something. Down went the grab. Seventy fahtoms of taut wire took the strain until it seemed it must part – then, slowly, slowly the grab surfaced. It held the entire forward end of the Comet: cockpit, flying controls, fire indicators, from the nose right back to the wings. Reluctantly the sea was giving up its ill-gotten harvest.

Two thousand tons of fuel oil had been expended on the operation and one hundred and forty thousand man hours of work. All the plane had not been recovered by any means, but enough to start the investigation into the cause of the accident, and work began on the mock-up.

Now it was April – and with it came a bitter blow. Another Comet, taking off from Rome's Ciampino Airport, this time for Cairo, was reported missing. It had disappeared with a crew of seven and fourteen passengers. The next day a few bodies and a little wreckage was found, but the depth to which the aircraft had dived into the sea made recovery of anything more virtually impossible.

Comets were immediately grounded and the certificate of airworthiness was withdrawn. Who can measure the heart-ache and fearful bewilderment of the Comet's manufacturers? Now the task of investigation of the first crashed Comet became desperate. Scientists worked on the mystery for impossible hours. The American Magazine *Aviation Weekly* described it as 'one of the most amazing pieces of scientific detective work ever done, the product of minute examination, probing analysis and deliberate deduction, held together through tortuous hours by British tenacity and national pride.' There are records of twenty-two days of court hearings, of the visits to Farnborough where the investigation was being carried out, and a gallery of photographs of all the exhibits. The dedication and sacrifice of all else to this sad and painful operation was a measure of the Comet manufacturers' concern and fierce pride in their reputation. At whatever cost the mystery had got to be solved.

There were experiments with gust-loads on the wings, pressurization and depressurization. Mock flights were tested above and beyond any normal flight endurance. Further salvaging work went on on the site of the first Comet's crash – and further wreckage was hauled up, including some major parts. At Farnborough the reconstruction progressed and now important factors were coming to light as to why the aircraft had failed. The tail unit, one of the last components to be recovered, revealed a startling piece of evidence: there were blue paint streaks on it and when the tail was fitted to the fuselage the streaks continued from front to back of the aircraft. Something had scraped along the side and had, possibly, torn off the tail. An analysis of the paint showed that it matched up with the paint on the seats. Also, a breathtaking find, a piece of carpet from the main cabin caught in the bent, twisted metal. It was beginning to add up. Hard on the heels of that discovery came the marks of scratches along the port wing. On examination of the scratches, paint matching up with the type of paint used on the fuselage, pointed an accusing finger at that part of the aircraft. Two conclusions were arrived at: the wing was intact when the fuselage failed, or the side of the fuselage had exploded through some great force. The next piece of wreckage that came in was a piece whose paint and broken edges matched with the markings on the fuselage.

The fitting of the puzzle pieces was reaching fever point when three days later something happened on the routine mock climb in the tank on the workshop floor. Without any warning the water which was being pumped into the tank failed to increase the pressure in the cabin. The gauge dropped to zero. On examination the incredible was found to have happened: the structure had failed due to fatigue at the corner of a window. Fatigue, after relatively few flight hours, was the last thing they had been looking for.

Now they looked for fatigue elsewhere, testing again and again, but no fault showed up on the pieces of wreckage recovered – but it couldn't be ruled out, there may have been weaknesses in the unrecovered parts lost at the bottom of the sea. It began to seem that they were never going to solve the mystery. July, and another search was made in the Mediterranean. Some pieces of wreckage were recovered in the trawls but nothing came to the surface that was of any great help. They went back to the water tank and the submerged mocked-up fuselage. Testing it under pressure it failed again at the corner of a window. It had to be that – but they needed proof.

On 8 August, an Italian fisherman fouled his nets on something – and brought up an enormous piece of metal. It turned out to be part of the centre fuselage with the frame of a window! In a corner of the window there was a telltale mark of fatigue.

Of course it was not possible to compare the wreckage of the two Comets, but there was evidence that linked them unmistakeably. Italian and British doctors were of one opinion that injuries sustained by victims in both disasters were the same – consistent with violent displacement and sudden contact with parts of the aircraft. Not only had a mock-up of the aircraft been made, but mock-ups of the passengers, and playing back the film taken at the moment when the pressure cabin broke left no doubts.

And so a chapter ended but a new one was to begin. What had happened could not be undone. It is a sad truth that so often a few must suffer for the good of the many. Valuable information had come to light, in fact, aviation history was made through the findings of this case. It enabled the manufacturers to 'design-out' the discovered weakness. The experience, though hardly bought, raised the newly developed Comets to the highest level of safety so that they set a standard for other manufacturers building jets and pressure cabins to follow. The tests that were carried out at Farnborough became the routine formula for all other aircraft manufacturers so that structural failure arising out of errors in design is now almost non-existent.

Wild Cats in the North Sea

Dorothy Clewes

Recently the sea has begun to give as well as take. Deep down, far below the sea-bed itself, through layers of different periods of chalk, and limestone, and rock strata millions of years old, deposits of gas and oil have been discovered. The search for this rich harvest from the sea has become a modern technological adventure.

It all started in Holland when a drilling rig, working in a field outside a village in a Northern province, struck one of the largest natural gas fields in the world. The discovery began the search for other fields as yet unsuspected and the search led to the North Sea.

Geophysical surveys were made on the structure of the sea-bed. These, sophisticated as they are, cannot determine the presence of oil or gas, but can point to under-surface strata where oil and gas may be found.

But the North Sea is hardly an ideal place to work in. It is storm-bedevilled, and winds of hurricane force can whip it into malevalent fury. On the other hand, the North Sea is a shallow sea. At its deepest it is about 650 feet and this reduces to 100 feet further South, to deepen again 50 or 60 feet as it approaches the English Channel. It has dangerous sand ridges, particularly, as all sailors know, the Dogger Bank. There are also similar smaller sand ridges lying off Norfolk – all to be respected. To anchor a drilling rig in these unpredictable waters is no mean feat. Accurate forecasts of the weather are essential.

1964 saw the first drilling in the North Sea. The first well is very much of a gamble – and a very expensive gamble it can be, so it is well-named 'wildcat'. It is rare that this first well will produce oil in sufficient quantity to warrant developing it, in which case the toolpushers, roustabouts, roughnecks – as the crew are variously called – must move on to drill for a strike in another area. In this way the oil company concerned can run into expenditure of millions of pounds before they strike it rich and the 'black gold' comes pouring in. So it is little wonder that oil goes on increasing in price, keeping up with the increasing price of the necessary equipment: observation chambers, decompression chambers, undersea workshops and laboratories for undersea water research, all taking place in depths impossible a few years ago. Shades of Beebe and Piccard, the great forerunners in this field!

Mobile rigs – which resemble man-made islands – were built in the U.S.A. as far back as 1949. Now there are almost two hundred of them operating all over the world, and more are being added to their number all the time. In the North Sea most of the drilling is done from these mobile rigs. They have legs that are lowered to the sea-bed, using a jacking system, and then they are firmly embedded there. The frame of the hull then jacks itself up, working much in the same way as a telescope works, until it is above wave height. They are massive structures which must house between fifty and seventy crew members, plus space to store away the heavy equipment and vast drilling pipe. Also, a necessity – space for a helicopter pad, for the helicopter is the work-horse, providing transport for the crew and all their requirements.

The sea-bed itself presents a problem as strong tidal

currents can move enormous quantities of sand, shifting it from around the legs supporting the platform, so that divers must be on hand to go down and see that all is well, and if necessary they must sandbag the legs to protect them from the erosion.

Work on a rig is very much a round-the-clock affair, the crew working in shifts of twelve hours on and twelve hours off, with compensating rest periods back on shore. And perhaps the most important members of the crew are the divers who must drill underwater to erect and complete the wells and thereafter maintain them. At first divers were trained for this very specialized work, turning them into underwater oilmen, but more recently it has been found more satisfactory to train oilmen and technicians to be divers. It is a new and fascinating world that they inhabit. It is not even necessary for them to dive down to their work base: a transfer bell – an underwater version of a lift – can take them down to the sea-bed, transfer them to the submarine workshop and to the site itself.

When a strike is made, then a network of submarine pipelines has to be laid to the shore treating plant. This treatment is necessary because of the impurities brought up from the underground reservoir, and once more this is the work of the intrepid diver, working away unseen and unsung in the dark depths.

The underwater oil and gas rush is on. It is doubtful whether the divers look on their work as romantic, but the finding of gas and oil in the North Sea has been called the most important discovery since James Watt watched the kettle lid bobbing over the steam and devised a way of saving power by the use of condenser and air-pump to drive the steam-engine. And so the discovery in Holland has triggered off what could be an economic and industrial revolution.

Electronic equipment and sophisticated engineering techniques, once worked out, have eliminated much human labour, but man can find no substitute for man, and in the quest for the underwater 'gold' who can tell what his inquisitive and probing mind will not find? Something at present beyond our understanding? Who knows?

Under Milk Wood

Dylan Thomas

Dylan Thomas in his much imitated way evokes an image for us in the dead souls of Captain Cat's remembered days. They speak to him from their deep drowned, sea-green sleep as he dreams in his bunk in the sea-shelled, ship-in-bottled, ship-shape best cabin of Schooner House.

First Drowned

Remember me, Captain?

Captain Cat

You're Dancing Williams!

First Drowned

I lost my step in Nantucket.

Second Drowned

Do you see me, Captain? – the white bone talking? I'm Tom-Fred the donkeyman . . . we shared the same girl once . . . her name was Mrs Probert.

Woman's Voice

Rosie Probert, thirty-three Duck Lane. Come on up, boys, I'm dead.

Third Drowned

Hold me, Captain. I'm Jonah Jarvis, come to a bad end, very enjoyable.

Fourth Drowned

Alfred Pomeroy Jones, sealawyer, born in Mumbles, sung

like a linnet, crowned you with a flagon, tattooed with mermaids, thirst like a dredger, died of blisters.

First Drowned

This skull at your earhole is

Fifth Drowned

Curly Bevan. Tell my auntie it was me that pawned the ormolu clock.

Captain Cat

Aye, aye, Curly.

Second Drowned

Tell my missus no I never.

Third Drowned

I never done what she said I never.

Fourth Drowned

Yes they did.

Fifth Drowned

And who brings coconuts and shawls and parrots to *my* Gwen now?

First Drowned

How's it above?

Second Drowned

Is there rum and laverbread?

Third Drowned

Bosoms and robins?

Fourth Drowned

Concertinas?

Fifth Drowned

Ebenezer's bell?

First Drowned

Fighting and onions?

Second Drowned

And sparrows and daisies?

Third Drowned

Tiddlers in a jamjar?

Fourth Drowned

Buttermilk and whippets?

Fifth Drowned

Rock-a-bye baby?

First Drowned

Washing on the line?

Second Drowned

And old girls in the snug?

Third Drowned

How's the tenors in Dowlais?

Fourth Drowned

Who milks the cows in Maesgwyn?

Fifth Drowned

When she smiles, is there dimples?

First Drowned

What's the smell of parsley?

Captain Cat

Oh, my dead dears!

Acknowledgements

The author and publishers would like to thank the following publishers and trustees for their kind permission to reproduce copyright materials:
Hutchinson & Co (Publishers) Ltd, for 'First Dip' from *Down to the Ships in the Sea* by Harry Grossett; The Bodley Head Ltd for 'Trapped' from *The Wonders of Salvage* by David Masters; Mrs Nicolete Gray and The Society of Authors, on behalf of the Laurence Binyon Estate, for *The Sirens* by Laurence Binyon; The Bodley Head Ltd for 'The Kingdom of the Helmet' from *Half Mile Down* by William Beebe; Sidgwick & Jackson Ltd for 'This is the Sea' from *A New Song about the Sea* by Edward Shanks; Longman Group Ltd for 'Two Miles Down in the Mediterranean' from *Seven Miles Down* by Jaques Piccard and Robert S Dietz; Harper & Brothers, New York (Hamish Hamilton Ltd, London) for 'Cave Diving' from *The Silent World* by Jacques Cousteau; Jarrolds Publishers (London) Ltd for 'Fish Talk' from *We Come from the Sea* by Hans Hass; George Allen & Unwin Ltd for 'Salt Pond Bay' from *Something Rich and Strange* by R E Schroeder; Hutchinson & Co (Publishers) Ltd for 'A Cache of Silver' from *Island Treasure* by Roland Morris; The Trustees for the Copyrights of the late Dylan Thomas and J M Dent & Sons Ltd for *Under Milk Wood* by Dylan Thomas; Simon & Schuster Inc for 'Jonah and the Whale' from *The Bible Designed to be Read as Literature* edited by Ernest Sutherland Bates. Special thanks are due to the generous material given by Mr P J Birtles, the Deputy Public Relations Manager of Hawker Siddley, used in 'Resurrection from the Sea', and Mr H F Wickham for his information on North Sea Oil and Gas.